A COMPREHENSIBLE UNIVERSE

The Interplay of Science and Theology

George V. Coyne
Michael Heller

A COMPREHENSIBLE UNIVERSE

The Interplay of Science and Theology

 Springer

George V. Coyne
Vatican Observatory
2017 E. Lee St.
Tucson AZ 85719
USA
gcoyne@as.arizona.edu

Michael Heller
ul. Powstancow Warszawy
13/94
33-110 Tarnow
Poland
mheller@wsd.tarnow.pl

ISBN 978-3-540-77624-6   ISBN 978-3-540-77626-0 (eBook)

SPIN 12161362

DOI 10.1007/978-3-540-77626-0

Library of Congress Control Number: 2008922391

© 2008 Springer-Verlag New York

The photographs on pp. 116–119 were taken with the Vatican Advanced Technology Telescope at the Mt. Graham International Observatory, Arizona, USA and are printed here with the kind permission of the Vatican Observatory Foundation.
Cover Image: Photo-composition of the William Herschel Telescope's dome and the star Sirius. By Nik Szymanek and Ian King, courtesy of the Isaac Newton Group of Telescopes, La Palma.

Typesetting and production: le-tex publishing services oHG, Leipzig, Germany
Cover design: eStudioCalamar S. L., F. Steinen-Broo, Girona, Spain

Printed on acid-free paper

9 8 7 6 5 4 3 2 1

springer.com

PREFACE

Every book has two lives: one before it is published and the other afterwards. In its life after publication, if the book is a success, it is studied by literary critics and finds its place in the history of literature. But its life before publication usually remains unknown; yet it is this part of the book's history that either leads it to success or consigns it to oblivion. The present book had its beginnings in a course of lectures delivered by one of the authors (M. H.) to students in the science departments at the Jagiellonian University in Cracow. The students played a vital role in shaping the structure of the course. Their questions and the lecturer's discussions with them helped to determine what topics would be treated in future sessions. Later on both authors met at the Vatican Astronomical Observatory in the Pontifical Palace amidst the bucolic surroundings of Castel Gandolfo where the papal summer residence is located. During long evenings, when the autumn winds went howling through the labyrinth of corridors and staircases in the palace, having completed their usual day of scientific research, they started working on the English version of the manuscript. While working together at Castel Gandolfo it was often easier and quicker for the co-authors to communicate via e-mail than to search for one another in the vastness of the palace. So when in due course they became separated by about half of the Earth's circumference – M.H. at his universities in Poland and G.V.C. at the Vatican Observatory's research institute in Arizona – they continued their work through e-mail as effectively as before.

In the meantime the Polish publisher *Proszynski i S-ka* expressed the desire to publish our work in Polish. The book was translated from English by Robert M. Sadowski and appeared in 2007 under the title *Pojmowalny Wszechświat*

(A Comprehensible Universe). The present English version differs from the Polish in a few respects. We have expanded the section on induction; we have added *Afterthoughts* at the end of the book and several bibliographical references. And so this little book enters its public life.

We express our gratitude to Abner Shimony for the important comments he made upon reading the manuscript and to Angela Lahee for her dedicated work at all stages of the production of this book.

December 2007
George V. Coyne
Michael Heller

CONTENTS

CONTENTS

CONTENTS

CONTENTS

INTRODUCTION

How is irrationality possible? We are asking this question not to provoke the reader. We think it is a serious question. It was Immanuel Kant who in his *Prolegomena to Any Future Metaphysics* asked the questions "How is pure mathematics possible?" and "How is the science of nature possible?". In Kant's time there was no doubt that pure mathematics and the natural sciences did exist. If they exist, they are possible. But how? How is it possible, that, when asking nature correct questions with the help of correct methods, we obtain the correct answers? Which conditions must the world satisfy, and which conditions should we ascribe to our cognition, that this procedure should be so marvelously effective? This was Kant's problem.

Irrationality does exist. To fight against facts would be the symptom of foolishness. But how is irrationality possible in the otherwise rational world?

A stone falls down. From among an infinite number of possible paths it will choose the one that is prescribed by Newton's law. It is true that elementary particles do not follow deterministic trajectories but even they are strictly subject to the laws of quantum physics. In the physical world nothing can exist that would be self-contradictory and exempt from mathematical regularities. Only we humans are free to be irrational. If somebody claims that "two times two makes a lamp", no catastrophe occurs, the universe continues to exist, and only that person, if he decides to act according to his claim, will experience the consequences of his peculiar metaphysics.

It is intriguing that we have a natural inclination to defend irrationality. One hears people saying that to be irrational is not just a human privilege. Computers can behave irrationally, but they always operate "in the framework of logic". They can execute a stupid program, for instance, they can translate a paper by Einstein into a sequence of meaningless symbols, but they cannot work without logic or against logic. If they are forced to do so by a programmer, they simply stop working.

We are free to be irrational. But this freedom has its price. We are living in the rational world. Something that is black will not become white only because I wish it to be so. The Earth will preserve its spherical shape even if everyone believes it to be flat. And if we stick to our irrational decisions, we will sooner or later pay a high price.

In this book we do not pursue the question of how irrationality is possible. We approach the problem from the other side. We rather explore the deep roots of – we do not hesitate to say – the Mystery of Rationality. For Einstein rationality was a matter of religion. In his essay on "Science and Religion" we read: "But science can only be created by those who are thoroughly imbued with the aspiration towards truth and understanding. This source of feeling, however, springs from the sphere of religion. To this there also belongs the faith in the possibility that the regulations valid for the world of existence are rational, that is, comprehensible to reason. I cannot conceive of a genuine scientist without that profound faith. The situation may be expressed by an image: Science without religion is lame, religion without science is blind."

Contrary to this conviction of Einstein, a popular view introduces a tension, if not a contradiction, between science and religion. In this view science is an incarnation of rationality, whereas religion belongs to the sphere of subjectivity based on irrational premises. In this book, we question such a perspective. It is true that the religious faith of many people exhibits some irrational features, and that authentic religion penetrates deeply subjective layers of the human personality (Whitehead wrote in *Religion in the Making* that "religion is what the individual does with his own solitariness"), but, in fact, rationality and religion are more deeply interconnected than one would be ready to admit at first glance. Rationality is a value, and embracing this value could be thought of as a religious act. That was Einstein's perspective, and that conviction determines the perspective adopted in the present book. In pursuing it we follow a historical approach. When following the evolutionary paths leading to the consolidation of the scientific method, and its function in disclosing the structure of the world, we shall see how closely the threads of reason and faith are interwoven.

This book consists of three parts. In Part One we begin our journey with a look at our distant past when the struggle with the world's rationality began. At that epoch, perhaps the greatest discovery of all times was made. Sometime during the 7th and 6th centuries BCE a few people in the Greek colonies on the coast of Asia Minor discovered that it was worth trying to understand the world with the use of reason alone without any assistance from myths and legends. There were two consequences of this audacious enterprise: the origin of Greek philosophy and the slow erosion of pagan religions. In some philosophical systems, to be sure, there appeared a god or a deity (Plato's demiurge or Aristotle's First Mover) but this deity was rather a "closure" or logical justification of their philosophical world view than the object of worship. Greek philosophy, especially its three great traditions: Platonic, Aristotelian and Archimedean, initiated the process of exploring the world's rational structure; rational – in the sense that this process discloses its secrets only when compelled by rational methods of investigation.

In Part Two we ponder on the "input of Christianity" to the process of exploring the rational structure of reality. Our culture has two great sources: Greek philosophy and the Judeo-Christian religion. No wonder new things happened when these two sources coalesced. After the initial, somewhat turbulent phase, a kind of synthesis was elaborated. The Christian God became the guarantor of rationality, but the Greek idea of rationality infiltrated the very core of Christian theology.

As the centuries passed, Greek rationality underwent further transformation. It was doubtlessly medieval Scholastic philosophy that constituted a link between the foundations laid in antiquity and the origin of modern science. If the Church Fathers saved Greek culture for us, the Middle Ages transmitted it to us. But it was by no means a passive transmission. The Greek concept of rationality had to go through all the intricacies and abstractions of Scholastic philosophy to finally prepare the ground for the emergence of the scientific method. Part Three deals with this process. There are many historical and philosophical books analyzing this period, and we do not want to compete with them. We make recourse to the history of science only to unravel a new plot prepared by the unpredictable logic of historical processes. And the plot is the following. The world is rational, in the sense that it can be rationally investigated. There are many possible methods of investigation, but when modern physics started the method of constructing mathematical models of various aspects of the world and of checking them experimentally, the progress of science became so rapid that it cannot be paralleled with anything else. This allows us to speak of the *mathematical rationality* of the world, or to say simply that the world is *mathematical*.

Here ends our analysis based on history and, in a closing chapter, we take a look at how the mathematical-empirical method works in contemporary physical theories (relativity theory, quantum mechanics and dynamical chaos theory) and research programs (the unification of physics and quantum gravity). When looking at these new incarnations of the world's rationality, the question arises: can everything be mathematized? The nature of the limits of the scientific method is certainly an important philosophical problem, but it also opens up new horizons. The principal tenet of rationality is that one is never allowed to cease asking questions if there remains something to be sought.

This book is not intended to be a scholarly work in the proper meaning of this word, i.e. an exhaustive monographic study of all aspects of the problem. It had its origin in a series of university lectures delivered by one of the authors (M. H.) to graduate and postgraduate students of mathematics, physics and other natural sciences who were not yet initiated into philosophical matters but who possessed the natural curiosity typical of their future professions. The intention of the authors is that this book should preserve its introductory character. Based on what a beginner-scientist, or simply an intelligent reader, is expected to know, it tries to open broader vistas. In this case, broader means also deeper – to the very roots of rationality.

PART I

~

THE DRAMA OF RATIONALITY

Why is our world comprehensible? This is a truly dramatic question. It is dramatic because it seems so obviously trivial that only a very few people dare to ask it. And if this question is asked, it turns out to be exceedingly difficult to answer. But quite independently of whether it is asked or not, all our success in understanding the world and all our technological progress depend crucially on this question.

One of the best ways to approach an answer to such a difficult question, or at least to come to a better understanding of it, is to look at the circumstances in which it was first asked. This happened in Greek antiquity. At the turn of the 6th and 5th century BCE great changes occurred in the history of human culture. Several audacious people tried to understand the world with no help from supernatural forces. So began the first ever conflict between the newly born rationality and religion. A slow erosion of mythical religions was an irreversible process. We discuss this in Chap. 1.

Very early on with the Pythagorean school it turned out that the comprehensibility of the world had something to do with mathematics. Greek thinkers faced the challenge. Plato did not doubt that beauty is an objective property. His demiurge, in creating the world, had no choice. He was obliged to choose the most perfect (i.e. the most symmetrical) geometric structures as models of physical reality. In the tradition initiated by Plato physics is an a priori science: to comprehend the world, it is enough to identify the most symmetrical mathematical forms. We discuss this in Chap. 2.

Aristotle's views were opposed to those of his master, Plato. Mathematics provides too simple a kind of knowledge to be able to cope with the richness

1

and proliferation of forms in the world. Mathematics studies only quantity and in the real world there are many qualities which remain beyond the reach of mathematical methods. Physics is the science that deals with causes, of which the final cause is the most important, and mathematics is unable to grasp a causal nexus. In the Aristotelian tradition physics is an a posteriori science. One should begin with sensual cognition, discover all sorts of causalities and, in this way, identify the nature of things which is the goal of physics. Mathematics has no major role to play in the sciences. We discuss this in Chap. 3.

For a long time Archimedes was considered as a "first among Platonists" but, in fact, he founded a third Greek tradition. His main difference with Plato was that, although he also dealt with mathematical structures, he did not select them a priori. He identified them by experimenting with simple mechanical contraptions (the lever, the balance, etc.). Once the "experimental situation" was put into a mathematical form he was able to deduce (to predict) the outcome of experiments, and then confirm them with the help of new measurements. Today we admire the Platonic and Aristotelian traditions, but we all have to study in the first chapters of our physics textbook exactly the same mathematical models that Archimedes so skillfully constructed. We discuss this in Chap. 4.

What happened at the beginning of science deserves thorough reflection. Sooner or later science becomes a problem for itself. And so, the philosophy of science is born. One of the greatest problems for the philosophy of science is to consider the very rationality of the world. We are back to the question: why is the world comprehensible? We do not answer this question but we try in Chap. 5 to show its highly non-trivial character and to examine it as fully as possible.

Chapter 1

~

DISCOVERY THAT THE WORLD
IS RATIONAL

1. THE GREAT MUTATION

*I*t seems that only geniuses are able to wonder about ordinary things. In fact, it is exactly because of this ability that they deserve to be called geniuses. The most ordinary urge is to try to explain why things are as they are. There exists an instinct, deeply rooted in the human nature, to explain things. This instinct very often leads to successful results. To it we owe, for instance, science and philosophy. Why is it then that so many things in the world, perhaps all of them, can be explained or understood? Why is the world comprehensible?

Albert Einstein expressed this wonder in a very clear way:

> *"The very fact that the totality of our sense experiences is such that by means of thinking it can be put in order, this fact is one which ... we shall never understand. One may say "the eternal mystery of the world is its comprehensibility".[1]*

There are some questions which, even if unanswered, are highly significant. "Why is the world comprehensible?" is one of these. Three elements are involved in it: (1) the world to be comprehended, (2) the human mind attempting to comprehend it, and (3) science as the means toward comprehension.

The question itself says something about the mutual relationship between these three elements owing to which the attempt to comprehend is, at least partially, successful. Time in its passage has had the kindness to transmit to us a remark ascribed to Democritus: "Rather would I find one simple causal explanation than conquer the throne of the Persians."[2] This remark testifies to the great mutation, so to speak, in the cultural genes of humanity that occurred during the 7th and 6th centuries BCE in the Greek colonies on the coast of Asia Minor, when a few audacious thinkers sought to comprehend the world with the help of their own mental capacities without any recourse to myths and legends.

This early discourse about nature was expressed in everyday language, the same language in which people communicated with one another and spoke about ordinary things. This certainly contributed to treating the world as a great "social organism" inhabited by various gods. These gods were supposedly equipped with a kind of higher intelligence, but they were led by passions and desires rather than by rational thinking in their rule over the world. However, this kind of irrational behavior of the gods did not exclude attempts to comprehend the world and its forces. On the contrary, the mystery of the world was a challenge which had to be, if not understood, then at least tamed. And that was the function of myths.

One of the oldest documents testifying to this human passion to tame the incomprehensible is the Babylonian poem *Enuma Elish*. The earliest version of this poem comes from the second millennium BCE, but there are reasons to believe that its roots go much farther back in time. An analysis of the story shows that the god Marduk, whose glory the document proclaims, took the place of an earlier deity, probably named Enlil. It turns out that the nasty and notorious habit of arbitrarily changing past history is not an invention of our times, since in later versions of the poem, coming from the period of the Assyrian domination, Marduk was replaced by Assur. "In this myth, the origin of the world is the result of a conflict between activity and inertia, order and chaos. In this conflict the first victory over inactivity is gained by authority alone; the second, the decisive victory, by authority combined with force."

The first decree of Marduk, as soon as he gained power, concerned the organization of the calendar. To this end he created the sun, the moon and the stars. The harmonious motions of heavenly bodies, synchronized with the seasonal changes on the Earth, were among the first factors that inspired early people to regard the world around them as something ordered, cosmos rather than chaos.

2. ORDER AND NECESSITY

Our reconstruction of the origin of rational thinking is largely based on hints and hypotheses, but historians agree that the first breach in the purely emotional approach to nature was brought about by a reflection on cyclic phenomena: the regular motions of the heavens, the seasons of the year, the regular flooding of rivers. It is not just coincidental that the first civilizations were born in the basins of great rivers. The motions of the heavenly bodies, however far away they were, seemed to dominate over everything; even the passions of the gods of Olympus were subject to their power and impartial preciseness. This led to the Greek idea of *fatum*, a blind necessity ruling the fates of gods and humans. This can be seen as an anticipation of the much later concept of the laws of nature.

Olaf Pedersen writes that "the new idea of an inherent necessity in nature arose among the Greeks and was never heard of in Egypt or Mesopotamia."[3] Moreover, the Greeks also invented a new "technical term" to denote this necessity. The Greek term *ananke* originally signified coercion or even torture. For instance, Herodotus tells the story of a criminal who was forced by police to confess his crimes under *ananke*. Gradually, the term was used by philosophers to denote "that strange something in nature which the phenomena are unable to resist."[4] Pedersen sees in this an instance of a more general process:

Throughout the centuries Greek philosophers pursued such experiments with a metaphorical language. The result was a vocabulary of technical terms the metaphorical origin of which went into oblivion in the course of the long process which gradually made the Greek world familiar with the new insight.[5]

Whitehead believes that:

… there can be no living science unless there is a widespread instinctive conviction in the existence of an order of things, and, in particular, of an order of nature[6]…

and he explains:

> *I have used the word instinctive advisedly. It does not matter what men say in words, so long as their activities are controlled by settled instincts. The words may ultimately destroy the instincts. But until this has occurred, words do not count.*[7]

Today we know that order can be very sophisticated. It can even be made up of chaos, and in the chaos itself there can be hidden regularities that are subject to mathematical analysis. Without an underlying order and its mathematical analysis we would be sentenced to a purely metaphorical language and, consequently, to an emotional relationship to nature. Some people practice such an approach to nature even today, but science mercilessly ignores their efforts.

3. THE FIRST CONFLICT BETWEEN REASON AND BELIEF

The beginnings of critical thinking inevitably influenced religious beliefs. In modern language we could say that the laicization of mythical religions was inevitable. Although the immediate crisis struck only rather narrow religious elites, its effects in the long range were enormous. This fact may serve as a warning against the claim that only mass processes count in history. A few generations of Greek thinkers created a world vision practically void of any religious elements, or at least with religious beliefs removed to the faraway margins of intellectual concerns. Early Christian writers were clearly aware of this process of the erosion of mythical religions. St. Augustine writes on Anaximenes that although he :

> *… did not deny that there were gods … yet he did not believe that air was made by them, but rather that they arose from air.*[8]

And Clement of Alexandria reports that Xenophanes openly ridiculed the gods of Olympus by writing that:

> *... if cattle and horses or lions had hands or were able to draw, ... horses would draw the form of their gods like horses, and cattle like cattle.*[9]

This process could be regarded as a first ever conflict between religion and science or, more strictly, between mythical religion and the beginnings of critical thinking.

This process was accompanied by another one, also pregnant with important consequences. In some philosophical systems there appeared a god or a deity. Such a god or deity was not an object of worship, but was considered rather as a sort of "ideal closure" of a given philosophical system. Such was Plato's demiurge who, in making the world out of "things that were in a state devoid of reason or measure,"[10] acted in accordance with pre-existing but atemporal ideas,[11] and Aristotle's First Cause or First Mover, who is "an eternal substance, which is unmoved and separated from all things that can be perceived by the senses."[12]

4. THE PRESENCE OF MYTH

When thinking about the origin of science and philosophy we often show a tendency to underestimate the value of myth. It was critical, rational thinking that replaced the irrational, mythical approach to unknown phenomena. We treat myth as a fable or a fairy tale invented by primitive people to recount to their sons and grandsons. However, the story has yet another aspect. There is something unknown, something that transcends our current knowledge and we want to comprehend it, but we are lacking adequate tools to do so. Therefore, we create a myth so that we might at least assimilate this unknown into the realm of our actions. It is true that science and philosophy have converted many myths into rational knowledge, but it is far from true that we no longer use myths. In some schools of contemporary philosophy the myth concept is still widely employed, but it has evolved into a kind of technical term. Myth in this sense refers to any belief or conviction that transcends our human experience and refers to a reality that evades any precise linguistic description. Since this reality cannot be precisely described in our language, it cannot enter into a logical nexus with any linguistic description of our experience. This does not mean that this reality cannot be experienced, but, if it is experienced, it transcends any logically organized linguistic description. Myths, in this sense, do not constitute "second-hand" knowledge. On the contrary, they frequently concern some of the most important elements of human life. Let us enumerate a few such myths.

A. The Myth of Value. We come to realize that some components or aspects of our experience are permeated with value, and so we grasp them as participating in a reality that transcends our experience. We are able to describe our ordinary experience in a more or less precise language. We can also try to describe our "experience of value" in some linguistic terms, but this description, being "mythical," is not precise but metaphorical. Because of this "lack of proportion" these two descriptions cannot enter into a logical interaction: they are logically incommensurable.

B. The Myth of Rationality. This myth reflects a conviction that our rational methods of investigating the world are not merely a *savoir vivre* of some eccentric people but reflect something that transcends us. The Myth of Rationality, like all myths, cannot be rationally established, because every argumentation presupposes the myth.

C. The Myth of Meaning. The world without this myth would be the world of individual instances and casual events. Although language and logic could try to compose a whole of such a world, there would be no reason for such an attempt. This myth is closely related to, or even a part of, the Myth of Value. Value without Meaning is meaningless, and Meaning without Value is pointless.[13]

Let us take a closer look at the Myth of Rationality. We can confidently say that the greatest discovery of the Greeks was the discovery that one's beliefs should be rationally argued for, that is to say that one should seek

> to solve as many problems as possible by an appeal to reason, i.e. to clear thought and experience, rather than by an appeal to emotions and passions.[14]

But the question immediately arises as to how to argue rationally that your beliefs should be argued for rationally? Karl Popper was a philosopher who fully understood the importance of this question. He wrote:

neither logical argument nor experience can establish the rationalist attitude; for only those who are ready to consider argument or experience, and who have therefore adopted this attitude already, will be impressed by them. That is to say, a rationalist attitude must be first adopted if any argument or experience is to be effective, and it cannot therefore be based upon argument or experience. So rationalism is necessarily far from comprehensive or self-contained.[15]

Why then should we not adopt irrationalism? Because when one confronts rationalism with irrationalism, one immediately sees that rationalism is a value. Therefore, "the choice before us is not simply an intellectual affair, or a matter of taste. It is a moral decision".[16] Indeed, the choice of value is the moral decision. Popper calls this kind of rationalism critical rationalism, the one "which recognizes the fact that the fundamental rationalist attitude results from an (at least tentative) act of faith – from faith in reason".[17]

Chapter 2

~

SHOULD THE ASTRONOMER LOOK
INTO THE SKY?

1. PLATO'S PHILOSOPHY OF PHYSICS

*E*ach reform of the state must contain a reform of education. Plato was
well aware of that. When in his *Republic* he was working out the project of education, he was certain that the ideal school should first of all
teach mathematics, in particular plane geometry. Glaucon, with whom Socrates
had a debate (in Plato's dialogues it is Socrates who presents the views of the
author), shared this opinion and suggested that astronomy should take the
second place.

"Anyone can see", said Glaucon, "that this subject [astronomy] forces the
mind to look upwards, away from this world of ours to higher things." "Anyone
except me, perhaps. I do not agree," answered Socrates. "Why not?"

> *Those intricate traceries in the sky are, no doubt, the loveliest and most per-*
> *fect of material things, but they are still part of the visible world, and there-*
> *fore fall far short of the true realities, the true movements, in the ideal world*
> *of numbers and geometrical figures which are responsible for these rotations.*
> *… So if we mean to study astronomy in a way which makes proper use of the*
> *soul's inborn intellect, we shall proceed as we do in geometry, working at*
> *mathematical problems, and not waste time observing the heavens.* [1]

Plato's strange proposal was the consequence of his philosophical views.
In creating his philosophical vision he was influenced by the Pythagorean

11

school. Two of Plato's teachers, Theodorus of Cyrene and Archytas of Tarent, were Pythagoreans, and Plato himself traveled to *Magna Graecia* (the Ionian coast of southern Italy) where this school was flourishing.

The followers of a half-mythical thinker, Pythagoras, created something between a philosophical school and a religious association. Their religious attitude was permeated by, so to speak, "experiencing the cosmos" in the light of their scientific achievements. We know their doctrine only from reports of later authors and to read it from our own viewpoint, full of our scientific sophistication, could become a trap which misleads us. When, for instance, the Pythagoreans of the early period believed that all natural objects are made of integral numbers and that it is the "number" that is the principle (*arche*) of the cosmos, they probably treated numbers as some people now treat atoms. And exactly because of this view the discovery that there exist irrational numbers was for them a shock;[2] such a shock was it that, as history records, they kept it secret and killed the traitor who revealed it.

The discovery of irrational numbers was the first ever global revolution in mathematics and it entailed a revolution in the understanding of the world. Integers ceased to be "atoms of the world" and became more abstract entities or "mathematical forms." Only as such were they elevated to the rank of a "principle of reality." This happened in Plato's system.

The astronomer should not waste his time by looking at the heavens since observational exploration of the world leads only to probable knowledge or opinion. Our senses can err and do so very often. True knowledge can only be obtained by deduction which is nothing other than an exploration of the world of ideas or eternal forms. To these two kinds of cognition (opinion and true knowledge) there correspond two kinds of beings:

> *first, the unchanging Form, ungenerated and indestructible … invisible and otherwise imperceptible; that, in fact, which thinking has for its object.*[3]

This is the world of ideas, the important part of which is inhabited by mathematical ideas.

> *Second is that which bears the same name and is like that Form; is sensible; is brought into existence; is perpetually in motion, coming to be in a certain*

place and again vanishing out of it; and is to be apprehended by belief involving perception.[4]

Material beings, perceptible by the senses, are only the shadows of the corresponding ideas. They even borrow their names from them; for instance, a cube made by an artisan is only an imperfect similitude of the Perfect Cube.

At this point, Plato's ontology meets Platonic aesthetics. For the Greeks, beauty was an almost physical property of bodies and, under the influence of the Pythagoreans, was identified with symmetry that could be represented by numbers with the help of various kinds of proportion. Later on this became known as the Great Theory of Aesthetics.

Plato had no doubt that the most beautiful (the most symmetrical) of all geometric solids is the sphere, and that the most perfect motion is uniform, circular motion. There is no need to observe the heavens to reach the conclusion that celestial bodies must follow such motions. In his *Republic* Plato even sketched a cosmological model starting from these principles and his disciple, Eudoxos of Knidos, improved this model and elaborated it mathematically.

In the Greek tradition, going back to the first Ionian philosophers, the world is composed of four elements: earth, water, fire and air. This time it was another of Plato's disciples, Theaetetus, who first did the geometric work and only later did Plato apply it to his theory of the "micro-world." Theaetetus proved that there are exactly five regular solids (the most symmetrical after the sphere): the cube, the tetrahedron or pyramid, the octahedron (eight equal surfaces), the dodecahedron (twelve equal surfaces) and the icosahedron (twenty equal surfaces). Then Plato used this classification to explain that earth is made of isosceles triangles, fire of pyramids, air of octahedrons, and water of icosahedrons. The shape of the remaining dodecahedron was ascribed by Plato to the Whole.[5]

2. PLATONIC INSPIRATIONS

Modern physicists, such as Heisenberg and von Weizsäcker, quite often see in Plato their own predecessor, and in Platonic philosophy a kind of archetype of the mathematical method of contemporary mathematical physics. The latter involves identifying the structure of the world with mathematical structures and it is exactly this identification that more often than not leads to extraordinary agreement between theoretical predictions and empirical results. More-over, without this "Platonic method" many important fields of research, for instance the whole of subatomic physics, would forever remain inaccessible to our scientific curiosity. This fascination with Plato's doctrine is reinforced by the fact that the most important laws of contemporary physics have the form of symmetries, although these are much more abstract than the forms exhib-ited by the Platonic solids. If we also remember that the creators of modern physics, both Galileo and Newton, remained under the strong influence of Platonic doctrine, we can understand better why many scientists regard mod-ern physics as an implementation of Plato's epistemological ideal.

However, the matter is not that simple. The astonishing agreement of some aspects of modern physics with Platonic intuitions should not screen out es-sential differences between Plato's proposals and the present scientific enter-prise. Mathematical structures with the help of which physicists model the world are indeed beautiful or, more precisely, they are regarded as beautiful by many physicists; but aesthetics is no longer regarded as a physical category. One cannot employ certain mathematical structures in physics only because they seem beautiful to even the most eminent physicists. Today the symme-tries of the Platonic solids seem to us less beautiful than, say, symmetries rep-resented by the groups $SU(2)$ or $SL(3,1)$, so often used in current theoretical research, and there is every expectation that there exist yet more beautiful structures which will in the future serve to model the world. And what if we are inclined to treat some mathematical structures as beautiful precisely be-cause they fit so well our empirical results?

Contemporary astronomers do not regard it to be a waste of time to look into the sky and to construct complicated instrumentation to do this in a more and more accurate way. They are fully aware of the fact that mathematics can be successfully applied to the world not in an *a priori* manner but in constant dialogue with the world which speaks to us in the language of observations and empirical measurements. However, and this is Plato's point, it is true that a well-chosen mathematical structure often discloses more information about the world than was contained in previously known empirical results; but this astonishing circumstance does not change the fact that it is experiment and observation which finally determine those mathematical structures most use-ful to model the structure of the world.

Chapter 3

~

SEVEN FIGHTERS AGAINST THEBES

1. THE WRATH OF ARISTOTLE

*A*ristotle devoted the last book of his *Metaphysics* (Book XIV) to a polemic against his master, Plato. Usually, Aristotle's style is scientifically emotionless, as if on purpose washed clean of any literary ornaments, so as to separate rational argumentation from any rhetoric. But in his polemic with Plato Aristotle allows his emotions to come to the surface. How otherwise is one to explain the following passage directed against Platonic doctrine:

> *All this is absurd, and conflicts both with itself and with probabilities, and we seem to see in it Simonides' "long rigmarole", for the long rigmarole comes into play, like those of slaves, when men have nothing sound to say.*[1]

Aristotle mercilessly ridicules the Pythagorean doctrine on number as an *arche* of the world:

> *There are seven vowels, the scale consists of seven strings, the Pleiades are seven, at seven animals lose their teeth (at least some do, though some do not), and the champions who fought against Thebes were seven. It is not because the number is the kind of number it is, that the champions were seven or the Pleiades consists of seven stars? Surely, the champions were seven because there were seven gates or for some other reason, and the Pleiad we*

15

count as seven, as we count the Bear twelve, while other people count more stars in both.[2]

It was not difficult for Aristotle to find out weak points in the Pythagorean-Platonic doctrine, naively applied to various situations. The essential difference between Aristotle's views and those of his former master was hidden in the very different conception of science. For Aristotle it was not important that there *were* seven champions fighting against Thebes but *why* there were seven champions. Scientific explanation of a phenomenon does not consist in finding its archetype in the world of ideas, but rather in identifying its cause. There were seven champions fighting against Thebes *because* this city had seven gates, and each champion was attacking one of them (or, perhaps, there was some other similar reason).

We instinctively sympathize with this view of Aristotle and we are inclined to think that it is based on common sense. This is exactly the point. Aristotle's philosophy of science is based on common sense but, as we shall see in the subsequent chapters, common sense proves often to be misleading as far as the interpretation of science is concerned.

2. MATHEMATICS AND PHYSICS

Aristotle was too good a thinker not to appreciate mathematics. In his writings one can find many attempts to use mathematics to describe some natural phenomena. The most interesting one is perhaps his attempt to mathematize motion (see the end of Book VII of *Physics*) and historians of mathematics usually feel obliged to devote a few pages to the mathematical contributions of his school. However, this does not change the fact that in Aristotle's view the role of mathematics in physics is only subsidiary and, as it were, accidental. Moreover, he thinks that too much of mathematics leads to a deviation of physics. Aristotle has several arguments supporting this view.

The most important of them is implied by his conception of natural sciences, in particular the conception of physics. In the first paragraph of the first book of *Physics* Aristotle argues that, since to understand a thing means to know its causes, it is clear that in order to understand nature we must discover its causes. And there are four fundamental causes in nature: material cause, that out of which a thing is made; formal cause, that owing to which a thing is of its own kind; efficient cause, that which produces a new activity; and final cause, that for the sake of which something is made. Mathematics can deal only with the formal aspect of things, but, strictly speaking, "number cannot be a formal cause".[3] Thus mathematics is insufficient to cope with natural phenomena.

In Aristotelian science final causes play the essential role. Natural phenomena are directed, so to speak, to reach their goals and the most efficient way to understand them is through their final causes. Aristotle's physics is *par excellence* a teleological science. Any goal is always something good that is to be attained but mathematical objects, being neither good nor evil, cannot serve as goals. In this respect they are useless and, consequently, they can play only a secondary role in physics.

Physics is a science of causes, and numbers cannot be causes in any sense. Aristotle makes this clear:

> *Number, then, whether it be number in general or the number which consists of abstract units, is neither the cause as agent, nor the matter, nor the ratio and form of things. Nor, of course, is it the final cause.*[4]

To draw the demarcation line between physics and mathematics was for Aristotle something much more than a mere methodological procedure undertaken for practical purposes. In his ontology reality is divided into certain "categories of beings" and the classification of sciences must be "true," i.e. it must reflect this ontological classification of beings. Mathematics and physics belong "from their very nature" to two different "epistemological planes" and the transfer of concepts and methods from one plane to another leads to mistakes in categories. Such a transfer was called by Aristotle *metabasis* and was considered by him a major error in doing science. He elaborated this doctrine against Plato's teaching on the nature of mathematics and its role in understanding the world.

Scientific practice is always stronger than artificial methodological rules. Aristotle himself had problems in this respect with the classical Greek sciences: harmony, optics and astronomy. They flourished and they made ample use of mathematics, without paying much attention to the Aristotelian *metabasis* principle. Aristotle treated them as a special kind of science, something between physics and mathematics, and his followers in the Middle Ages called them *scientiae mediae*.

Besides these theoretical arguments Aristotle also had in view some more practical ones. He believed that the real world is too rich, too complex in its variety of forms, to put it into rigid and overly simple mathematical structures. "But how are the attributes – white and sweet and hot – numbers?" he asks[5]. Or in another place Aristotle says that one cannot require mathematical preciseness in all cases; "it can be required only in the case of non-material things. Therefore, this method is not suitable for the natural sciences, since everything in nature is material".[6] Of course, we should remember that for Aristotle the words "all nature is material" mean something radically different than what they mean today for us. Our concept of matter is of a much later origin, whereas for Aristotle the term "matter" was a synonym of the term "material cause" and denoted pure potentiality to be actualized by incoming forms.

It goes without saying that Aristotle's views on the "incommensurability" of mathematics and physics were pregnant with consequences for the further development of science. It would be difficult to point out another doctrine in the history of science that would block progress for so many centuries. What was it that blinded such a great thinker? Was it his own philosophy? It has certainly supplied false arguments. However, we should look at these views of Aristotle with some tolerance, remembering that his doctrine was born in the polemics with Plato and the Pythagoreans, and to a common sense thinker such as Aristotle their doctrine had to look crazy. This is a good lesson for us: the logic of reality quite often surpasses our common sense to such a degree that it might seem crazy.

3. THE NATURE OF MATHEMATICS

There is yet another reason for Aristotle's defeat. The mathematics which he had at his disposal was still in its infancy. If one believes that the subject matter of mathematical research is "surfaces, and volumes, lines and points", although they should not be treated "as limits of physical bodies[7]," then, if mathematics remains inadequate to the task, one must switch in physics, sooner or later, to purely qualitative analyses.

The problem of the mutual relations between mathematics and physics is strictly connected with the question concerning the nature of mathematics. It would seem that the view on the nature of mathematics should determine the view on the relations between mathematics and physics, but the history of science clearly teaches us that both views condition each other, and that it is often impossible to tell which of them has more influence on the entire philosophical context. This applies in the case of Aristotle. According to him mathematics is an abstraction from material things. Mathematical objects do not exist without material things. There are material things that, besides qualitative properties, possess a "substantial nucleus" which cannot be reduced to quantity. In this respect, the Aristotelian system seems to be logically consistent. And this turned out to be dangerous. This "logical closure" of the system became a trap for many generations of thinkers. It cannot serve as the only criterion of truth.

Many people look with favor at the Aristotelian doctrine concerning the nature of mathematics because they confound the question of origin with the question of ontology. As far as the question of origin is concerned, Aristotle was probably right: we indeed form our mathematical concepts by abstracting from the qualitative aspects of real objects. But how we discover things need not coincide with how things are "in themselves."

In the dispute between Aristotle and Plato the question why the world is mathematical took a dramatic turn. During this dispute different standpoints crystallized and entered into the complex net of various philosophical doctrines. Although future generations will enrich the scene with new subtleties and new opinions, the polemics between Aristotle and Plato will always remain a kind of a reference point. After many centuries a third party, in addition to the followers of Aristotle and of Plato, will appear. This party will not openly take a stand in the dispute, but the very fact of its existence will often influence both the formulation of the problem and the direction of further inquiry. This third party will come about through the progress realized by those sciences that make ample use of mathematics as a tool in their investigations. Sometimes, instead of asking philosophers, it is better to look at what happens in the sciences themselves.

Chapter 4

~

HOW TO COUNT THE GRAINS OF SAND

1. THE NUMBER THAT IS GREATER THAN ANY OTHER NUMBER

There are some, King Galen, who think that the number of grains of sand is infinite in multitude; and I mean by sand not only that which exists about Syracuse and the rest of Sicily but also that which is found in every region whether inhabited or uninhabited.

These are the opening sentences of the short treatise by Archimedes entitled *The Sand Reckoner.*[1] In what follows Archimedes explains to King Galen that there are also people who believe that, although the number of grains of sand that could fill the entire Earth is not infinite, it is so huge that it surpasses any magnitude that could be named. And now, he, Archimedes, shall do something extraordinary. He shall construct a number which is greater than the number of grains of sand with which one could fill not only the Earth, but also the universe. And not only the universe as we usually imagine it, with the Earth in the middle, covered with the sphere of fixed stars, but also the universe of Aristarchus of Samos who claims that the Earth is not at rest but encircles the sun, and that the ratio of the radius of the circle along which the Earth goes around the sun to the distance to fixed stars is like the center of a circle to its radius. Archimedes goes on to clarify that the last sentence is only a metaphor since if understood literally it would be nonsense. It is intended only to express the immensity of Aristarchus' universe. Even if one were to fill

21

such a universe with sand, there is a number greater than the number of these sand grains.

This introduction is followed by the technical part of the treatise. Archimedes carefully enumerates his assumptions, formulates theorems, constructs proofs, and comments on the conclusions. Some historians of science say that the problem that Archimedes had to face could be reduced to the fact that the Greeks had no digital system which could easily deal with great numbers. Even if this is so, Archimedes' work leads to the non-trivial problem in number theory: is there a number greater than any number that could be thought of? The sand metaphor was not only for Archimedes a literary device but it also suggested, in a subtle way, at least two interesting philosophical problems: how to use mathematics in investigating the world? And: is mathematical infinity somehow realized in the real world?

Archimedes was first of all a mathematician, – some historians of science say the greatest mathematician of antiquity, and perhaps even one of the greatest mathematicians of all time.[2] And what about Euclid? Comparing geniuses is a risky enterprise, but one could say that Euclid was mainly the one who classified, doubtlessly in an ingenious way, all of Greek geometry (it is difficult to reconstruct what was his original contribution and what he composed out of existing elements), whereas Archimedes was an incomparable discoverer. It is enough to mention that his method of computing the volumes of various solids (the so-called "exhaustion method"), even if it cannot be regarded as a "simplified integration" (Archimedes lacked the concept of convergence), it certainly paved the way to the discovery of calculus many centuries later.

2. THE METHOD

Later generations, up to the European Renaissance, were not able to comprehend the scientific achievements of the mathematician of Syracuse, and remembered him mainly as an inventor of some ingenious mechanical contraptions and their surprising applications. A legend about how Archimedes defeated the Roman fleet by burning its ships with the help of concave mirrors that concentrated the sun's rays on them brought him more fame than his real scientific achievements. And his words to King Hieron: "Give me a place to stand and I shall move the Earth!" became proverbial. Then there is the story that to convince the king Archimedes moved a heavily loaded vessel with the help of a system of concatenated wheels. Cicero tells us in one of his writings that he himself saw a model of the planetary system constructed by Archimedes that functioned so precisely that it was possible with it to predict eclipses of the sun and the moon. Archimedes was undoubtedly also a genius at inventing and constructing various scientific instruments.

In this respect those who came after Archimedes underestimated him. They treated his constructions as interesting gadgets, whereas in fact they were true scientific instruments. By using some of them Archimedes performed many experiments in the field of statics and hydrostatics. He was gifted enough to combine his manual talents with his mathematical genius. There is no doubt that experiment and observation constituted a starting point for him, but it had to be experiment and observation of a very special kind. Only those properties of bodies that could be expressed in numbers attracted his attention. For instance, two equal weights, suspended on the arms of a balance, remain in equilibrium only if they are at equal distances from the support point. Note that this is not a direct report of what is experienced. There is no reference to any particular weights or to any particular balance. Thus the above statements contain an element of generalization, as well as of idealization: any experiment gives a result only within the margin of the measurement error; but Archimedes treats his result simply as valid.

The result of an experiment prepared in this way can be expressed in mathematical language or, as we would say today, in the language of mathematical equations. But the historical truth is that equations began to be used only in the sixteenth century. Archimedes should be admired all the more because he was able to think mathematically and perform mathematical calculations without the help of a developed symbolic language. Mathematical equations, which outsiders usually so abhor, in fact greatly facilitate the work of mathematicians.

Once an experimental result has been expressed in mathematical language, the rest follows from the consequences of logic. It works through mathematical deduction. In this way a mathematical model of a given physical phenomenon is created. Some of the conclusions obtained by mathematical deduction can be experimentally checked. If new experiments confirm theoretical predictions the model gains its credibility. If this is not the case, we must either reject the model or, what is more often the case, change some of its elements, and begin the investigation process from the beginning.

It goes without saying that expressions such as "model of a physical phenomenon" or "gaining in credibility" are borrowed from dictionaries of our present methodology, but the content to which these expressions refer was significantly present in the scientific work of Archimedes.

3. THREE GREAT TRADITIONS

In Chap. 2 we saw that Plato also applied mathematics to his investigations of the world. But Plato and Archimedes had different approaches. Plato made *a priori* choices of some mathematical structures because they were beautiful. The strategy of Archimedes was, on the contrary, *a posteriori*. Through observation and experiment it was nature herself which suggested the correct mathematical structure. The Archimedean approach does not impose our ideas of beauty on nature, but tries to discover those mathematical patterns according to which nature operates.

We admire Plato and the intuitions that prompted him to believe that there are underlying symmetries that govern the world of appearances. Contemporary physicists share this view. But if we treated his metaphoric enunciations literally it would be difficult to defend him against the objection of naivety. From our present perspective the achievements of Archimedes in physics are rather modest. He simply created the beginnings of statics and hydrostatics; but his results in these domains are repeated today in all elementary textbooks of physics. And it was Archimedes, and not Plato, who was entitled to jump naked out of his bath and shout: *Eureka – I have discovered.*

Today historians of science distinguish three great scientific traditions in antiquity: the Platonic tradition, where mathematics is used *a priori* to physical investigations; the Aristotelian tradition, which consists of qualitative, essentially non-mathematical physics; and the Archimedean tradition, where mathematics is applied *a posteriori* to physical research.[3] However, until not long ago historians of science regarded Archimedes as an eminent Platonist. Strange as it may seem, Archimedes himself was at least partially responsible for that. According to the Greek custom (so splendidly implemented by Euclid), he always tried to present his results in a ready, quasi-axiomatic form without disclosing the laborious path that led to them. Only the detailed analysis of his writings allows one to reconstruct his route to his scientific discoveries. Moreover, one might also suspect that many of his purely mathematical achievements were suggested to him by experiments he performed. That suspicion became a certainty with the discovery in 1906 by a Danish scholar, Johan Ludwig Heiberg, of a previously unknown work by Archimedes. The words were very faintly discernible under the script of a medieval prayer book written on a much older manuscript. By using advanced chemical and electronic methods it was possible to recover the older text. The work is dedicated to Eratosthenes, to whom Archimedes explains his method of producing mathematical theorems:

I thought it correct to write out for you and explain in detail in the same book the peculiarity of a certain method, by which it will be possible for you to get a start to enable you to investigate some of the problems in mathematics by means of mechanics. The procedure is, I am persuaded, no less useful even for the proof of the theorems themselves; for certain things first became clear to me by a mechanical method, although they had to be demonstrated by geometry afterwards because their investigation by the said method did not furnish an actual demonstration. [4]

The story about the death of Archimedes has become symbolic of mathematicians:

In the general massacre which followed the capture of Syracuse by Marcellus in 212 B.C., Archimedes was so intent upon a mathematical diagram that he took no notice, and when ordered by a soldier to attend the victorious general, he refused until he should have solved his problem, whereupon he was slain by the enraged soldier. [5]

It is true that in the centuries to come the Platonic and Aristotelian traditions were overwhelmingly dominant, but the Archimedean tradition never completely faded away. It was present in many astronomical investigations and in Arabic contributions to science, and it was splendidly revived in the work of Galileo and Kepler.

In considering Archimedes' thoughts on the grains of sand, so numerous that the universe is too small to contain them, one can see the ambition of a thinker who wants to force the Great Problem to obey the rules of logic. St. Augustine is reported to have said that to understand God is like trying to place the ocean in a little hole made in the sand on the seashore. Both of them, Archimedes and St. Augustine, belonged to a rare generation of Sand-Reckoners.

Chapter 5

~

IS THE WORLD RATIONAL?

1. QUESTIONS ABOUT RATIONALITY

*I*n the previous chapters we have contemplated the origins of scientific discourse and we have seen that, almost from the very beginning, it was accompanied by a methodological reflection on what was being carried out. Three great traditions: Platonic, Aristotelian and Archimedean, have initiated different ways of thinking about nature and our efforts to resolve its mysteries. In this chapter, in order to understand the roots of the scientific enterprise, we will look at these important early achievements from our present perspective. The breathtaking successes of modern science often make us forget the assumptions on which these successes are based. Nowhere is this more clearly seen than in the beginnings of science.

Science has two properties that seem to be distinctive: (1) its ability to generate new problems, and (2) its ability to invent new methods to solve these problems. After a moment of reflection we can add a third item to the above list: (3) science is very *effective* at both generating new problems and inventing new methods to solve them. This effectiveness can be seen in the way that predictions following from scientific theories are correct more often than not, and also in the innumerable technological applications deriving from science that are constantly changing our everyday lives. We can also discern a certain gradation in this invent-and-solve method of science. Old problems generate new ones which in turn inspire scientists to invent new methods with the help of which new problems can be solved. This is the first level or grade. But sooner or later, science becomes a problem for itself, the problem that has to be solved. This is the second level, also called the meta-level, and to explore it is a task for the philosophy of science. One of the greatest problems that arises on this level is the problem of the *rationality of science*. It assumes several forms. One can ask about the rationality of human cognition that leads to scientific knowledge. One can ask about its essential properties. One can ask many other questions, for instance: how is one to distinguish rational forms of human activity from its irrational forms? In what sense is science rational? Why should we be rational in doing science and in other areas of our lives as well? Is the evolution of science rational, i.e. has this evolution some inherent "internal" logic or is it subject to purely contingent "external" factors such as the psychology of scientists or the social and economic conditions in which they live and work? Finally, one could ask about the conditions the universe must satisfy in order to be the subject matter of rational inquiry. In what follows this last problem will be called the *problem of the rationality of the world*. It constitutes the main topic of the present chapter, if not of the entire book. Since we are asking here about the conditions the world must satisfy to render science possible, it seems suitable to qualify this rationality as the *ontological* type of rationality. We should emphasize that we call the world rational not because it is endowed with some cognitive powers (we do not believe that it is), but because it can be investigated by humans with the help of their rational methods.

The problem of the rationality of the world is a part of the problem science creates for itself. If it were not for science's enormous successes, the problem of the world's rationality would probably never have arisen.

2. OUR PRELIMINARY HYPOTHESIS

There exist divergent opinions concerning the criteria any rational discourse should satisfy, but the problem of the ontological rationality of the world excites especially heated discussions. It is interesting to note that opponents of the world's rationality are most often recruited from among philosophers, whereas for many physicists it seems rather obvious that we should ascribe properties to the world that make it amenable to rational investigation. As an example let us quote the famous passage from Einstein's *Physics and Reality*:

> *The very fact that the totality of our sense experiences is such that by means of thinking (operations with concepts, and the creation and use of definite functional relations between them, and the coordination of sense experiences to these concepts) it can be put in order, this fact is one which [...] we shall*

never understand. One may say that the eternal mystery of the world is its comprehensibility.[1]

The mystery, of which Einstein is speaking here, is seminally present in our everyday cognition of the world, but it reveals itself clearly only in the sciences. The word "comprehensibility" refers not only to the fact the world *can* be comprehended by us, but also to the fact that it actually, and to a large extent, *is* comprehended by us. The very act of doing science creates in the researcher a firm conviction of entering into cognitive contact with something that surpasses her or him and that reveals its mysteries in response to the hard work entailed by the scientific method. The investigation of the world is not easy, but possible. Moreover, the history of humankind is a witness to its success.

We are now going to formulate our preliminary hypothesis: it asserts that the world has a certain property owing to which it can be successfully investigated by us. We shall call it the *hypothesis of the rationality of the world* (or simply the *rationality of the world*, hoping that we shall not be misunderstood). We treat it as a starting point for our further analyses. The fact that we call this formulation a *hypothesis* is not intended to mean that it has weak arguments on its behalf; on the contrary, we believe that right from the beginning it is very well founded. We wish only to stress that we are not dogmatic in this respect, and that as our analysis progresses our hypothesis will be more and more strongly corroborated.

3. AGAINST RATIONALITY

The arguments against our initial hypothesis can be reduced to two objections. The first of these is to maintain that the world in itself is not rational, i.e. that it has no property owing to which we are able to investigate it, but rather it is our activity that introduces, or imposes, on the world an apparent order. In other words, it is not the rationality of the world that is the necessary condition of its successful investigation, but rather the rationality of our process of investigating; ultimately, it is human rationality, and not the rationality of the world itself, which is at stake. This argument is often quoted by both empiricist philosophers and thinkers with a background in the humanities. In the latter case, the argument often assumes the form of a statement that it is the human being who *projects* rationality onto the world.

One should admit that this argument looks attractive. In many instances we indeed introduce an order where there was no order. For example, we often rationalize our own not necessarily rational conduct. However, to find examples outside the field of psychology is much more difficult. We would be inclined to guess that, outside the sphere of our subjective life, it is only at the level of description that we are able to introduce order where it does not exist. For instance, despite the fact that the economy in communist countries was exceptionally irrational, communist propaganda presented it as very successful. However, this was purely a verbal kind of rationalization (on the level of description only) which in the end did not prevent communist economies from suffering catastrophic collapse.

An example of the opposite is provided by mathematical models of modern physics. Mathematical models used in physics not only *describe* some aspects of the world, but they also imitate them in some sense; or, more precisely, they model them, i.e. they function in a way similar to the investigated aspects of the world. In this sense mathematics is something more than just the language of physics. Some metaphorically say that it is "the stuff out of which physics is made." How should one understand this? A model is a mathematical structure, and within this mathematical structure one can perform various mathematical operations: solve equations, investigate the behavior of curves, deduce conclusions from mathematical premises, etc. If the model is successful in physics, then many of these operations are in strict correspondence with the actual operations of the investigated aspect of the world. For instance, by analyzing solutions of a certain system of equations we are able to predict future states of the system, and by studying curves on a certain space we can reconstruct the motions of stars or planets or other bodies. Of course, it is possible to construct a false model of a given phenomenon (the history of science provides a host of examples). Such a model predicts effects that do not occur in the world, and if this is so, the model is rejected. Here we come to the crucial point. You can easily construct a false model, but you cannot construct a mathematical model of something that is irrational. Irrationality introduces into the model contradictory elements, and contradictions destroy the model.

Another argument against the rationality of the world derives from evolutionary biology. There is nothing strange, some philosophers say, in the fact that we see the world as rational, since our rationality was produced by natural selection in a long evolutionary process. In the world we discover only what the world has implanted in us. This seemingly very convincing argument, in fact, does not destroy but strongly supports our hypothesis of the rationality of the world. If it was natural selection, i.e. ultimately mechanisms of adaptation to the world, that have enforced on our species a rational attitude to reality, then we must ascribe to the world a property (or a collection of properties) owing to which this enforcing process was possible. And this is exactly what we have called the rationality of the world.

4. THE KANT EFFECT

Let us continue the same line of reasoning. The human brain is a product of biological evolution and, as such, undoubtedly a part of the natural world. Thus the rationality of the world contains in itself the rationality of the human brain. The cosmic evolutionary process equipped the human species with self-consciousness, that is to say, the awareness of its own consciousness. And so, within the rationality of the world a new quality was born: the long process of thinking was initiated. After many millennia this process has led to the origin of science. The process of the evolution of human culture should be naturally regarded as a continuation of biological and even cosmic evolution, and consequently we could say that cosmic evolution, in its human expression, underwent enormous acceleration. Owing to this acceleration human rationality has become so rich and so autonomous that we are now inclined to treat it as independent of the rationality of the world.

We are fully entitled to think that, because of scientific progress, we continuously improve our understanding of the world and its workings. In this understanding, elements of our own rationality cooperate with elements of the rationality of the world. To some extent, when we investigate the world, we investigate our own rationality. This effect, doubtlessly existing in our cognition, could be called the *Kant Effect*. As is well known, it was Kant who believed that when knowing the world through our senses, we learn about the structure of our sensory equipment (about our sensory categories, in Kant's parlance) rather than about the structure of the "external world." The Kant Effect certainly exists, but it is not as important as Kant believed. His way of thinking was metaphysical; our way of thinking should be evolutionary. Even if we learn something about ourselves, we learn something about the world that has engendered us. Evolution has created this marvelous feedback between the rationality of the world and our own rationality.

PART II

~

THE INPUT OF CHRISTIANITY

Our culture was born from the encounter of Greek philosophy with the Judeo-Christian religion. This encounter has also shaped our present sense that rational Christianity was not directly interested in investigating the world. It assimilated Greek rationality mainly because it had to construct its own theology on Greek concepts and language.

This assimilation was far from being only passive. The Christian God became the guarantor of rationality but, on the other hand, the Greek idea of rationality has infiltrated the very concept of God. The God of the philosophers, understood as a "closure of the world," has became a God of religion and worship.

Cooperation and conflict are the very essence of the relation between Christianity and the sciences. Cooperation, since both Christianity and science have their roots in Greek rationality; conflict because the fundamental Christian claim that God has entered into human history goes beyond the scientific method. In the eyes of Christian theology the world's rationality is but God's Logos immanent in the world (see Chap. 6).

Christian theology was born in the period of the Church Fathers not as a product of intellectual curiosity but out of the need to practice and preach the new religion. There were two choices: either to be happy with common-sense tenets, or to undertake a process of rational reflection, and this meant using the resources of Greek philosophy. The first possibility was resignation rather than a choice. Two great personalities played an important role in this process: Origen paved the way, and Augustine established the standards for future generations (see Chap. 7).

It is interesting to follow the laborious adaptations of the Greek model of the world to the needs of Christian theology. The greatest challenge was the biblical doctrine of creation. Early Christian writers, while not abandoning the original biblical meaning, did not, on the other hand, resist the Greek instinct to seek the "mechanisms" of the world's origin, and they elaborated the doctrine of creation ex nihilo. It was also a reaction against Gnostic teaching that the "principle of evil" is inherent and resident in matter.

The doctrine of creation, together with the dogma of the Incarnation, has introduced a new understanding of time. The idea of closed time, the cyclic repetition of history, common in antiquity, had to be rejected. The history of the world lasts from its beginning, through the coming of Christ, until its completion. It is from this dogmatic stance that we have inherited the idea of linear time.

In the Patristic period there were three attitudes to the world: the pagan attitude of contemplating nature as a manifestation of the deity; the Gnostic attitude of regarding the world as a product of evil forces; and a Platonic attitude distinguishing between the transcendental world of eternal forms and their imperfect replications in the material world. Christian theology adopted the Platonic attitude and so saved rationality for our culture.

If Christian antiquity saved Greek culture and Greek science for us then it must be said that the Middle Ages transmitted them to our times (Chap. 8). But it was not a passive transmission. In this process the Greek concept of rationality was further strengthened and transformed. From this point of view medieval Scholasticism can be seen as an exercise in defining concepts and abstract reasoning, under the strict surveillance of logic, which was developed for this purpose. The medieval art of disputation had in itself strong elements of what is nowadays called the philosophy of language. It is hard to imagine the invention of operational definitions and the origin of mathematical reasoning as applied to natural phenomena without this preparatory work.

In the medieval form of metaphysics the Christian God became the "closure" of a philosophical system. The God of religion was identified with the God of metaphysics. Neither before nor afterwards did the idea of rationality have such strong support.

Medieval Scholasticism not only contributed to the origin of the new method of the empirical sciences, but it also had strong implications for the content of science and philosophy. For instance, medieval disputations concerning God's omnipotence gradually evolved into the modern concept of "laws of nature." What God can and cannot do has a clear bearing on what is implemented in the created world. Constraints on God's power became constraints on the functioning of the world, and this lead directly to the concept of "laws of nature."

Chapter 6

CHRISTIANITY ON THE SCENE

1. GREEK PHILOSOPHY AND THE BIBLICAL TRADITION

*T*he encounter of Greek philosophy with Judaism and Christianity was of unprecedented importance in the intellectual evolution of humanity. As we have seen in Chaps. 1 to 3, the interpretation of the concept of the deity as a closure of the "system of the world" played a significant role in the Greek idea of rationality. However, this "philosophical monotheism" had little in common with religious monotheism. The latter was at that time an exceptional phenomenon; it was cultivated only by the Jewish nation. The deity of the Greek philosophers was necessary to cause motion and to justify order in the cosmos; the Jewish God was acting in the history of the Jewish people and led them to Messianic fulfillment.

A conviction, quite common among theologians, that cosmology was only a lesser addition to the Old Testament doctrine, even if true with regard to earlier periods, requires revision as far as later centuries are concerned. In the theology of the Psalms references to cosmology are as frequent as those to history. In the Sapiential literature, Wisdom permeates the world and mediates between the cosmos and God. In the Book of Sirach Wisdom speaks of herself:

> *From the mouth of the Most High I came forth,*
> *and mistlike covered the earth [...]*
> *The vault of heaven I encompassed alone,*
> *through the deep abyss I wandered.*
> *Over the waves of the sea, over all the land,*
> *over every people and nation I held sway (Sir 24, vv 3–6).*

Both the Book of Sirach and that of Proverbs show a deep relationship between human integrity, the source of which is Wisdom, and the cosmic order of creation.

Wisdom in the later books has acquired more Hellenistic properties. In the Book of Wisdom (originally written in Greek!) she assumed something of the Stoic *logos* (or *pneuma*). As far as we know, the word *logos* was first used by Heraclitus of Ephesus to denote a rational principle ordering the world. In the Stoic doctrine *logos* was God, Nature and Rationality, all at the same time, a kind of substance present in everything. Human nature participates in *logos* and the ideal of the human life is to live in consonance with it. The author of the Book of Wisdom is fully aware of this:

> *Beyond health and comeliness I loved her [Wisdom],*
> *And I chose to have her rather than the light,*
> *because the splendor of her never yields to sleep.*
> *Yet all good things together came to me in her company,*
> *and countless riches at her hands;*
> *And I rejoiced in them all, because Wisdom is their leader,*
> *though I had not known that she is the mother of these (Wisdom, 7, vv 10–12).*

And to live in consonance with Wisdom means to penetrate the cosmic order:

> *For he [God] is the guide of Wisdom*
> *and the director of the wise [...]*
> *For he gave me sound knowledge of existing things,*
> *that I might know the organization of the universe*
> *and the force of its elements.*
> *The beginning and the end and the midpoint of times,*
> *the changes in the sun's course*
> *and the variations of the seasons [...]*
> *Such things as are hidden I learned, and such as are plain;*
> *for Wisdom, the artificer of all, taught me (Wisdom, 7, vv 15–22).*

A similar style is apparent in Jewish apocalyptic literature. In the prophetic writings stress was laid on obedience to Revelation, whereas in the writings of the apocalyptic tradition (which proliferated in the Hellenistic period) emphasis was placed on understanding the cosmic order. A part of this understanding consisted in a conviction that the cosmic order reaches beyond the borders of earthly life. This idea finds its continuation in the Book of Maccabees.

When speaking about the relationship between Judaism and Greek culture one should not forget Philo of Alexandria (about 20 BCE to 45 CE), a Jew educated in Greek philosophy. He interpreted the Jewish religion in terms of Platonic and Neoplatonic doctrines, and became a bridge between these two worlds.

If it is true that European culture was born from the encounter of Greek philosophy with biblical tradition, we are very close to its birthplace. Only one more factor is needed, the appearance of Christianity.

2. FOOLISHNESS TO THE GREEKS

Christianity had started as a sect within the Jewish religion, but very soon, within the course of a few generations, it embraced the whole of the ancient world. This phenomenon had many causes; we shall mention only two of them.

First, there was the universal character of Christian doctrine. The juridical interpretation of the Old Testament required that anyone who wanted to join the Old Testament religion had to become a Jew. Christianity, due mainly to the teachings of St. Paul, very soon stepped away from this tradition. Second, the development of Christianity was a process which fitted very well into the pattern of processes analyzed by us in the previous chapters. Although the erosion of mythical religions will last for a few more centuries, already at that time it was an irreversible process. Decaying pagan religions left an empty space. Philosophical doctrines were accessible only to the elite, and this was not enough. The God of the philosophers is an hypothesis, but one must build one's life on something more solid. Christianity was not only a religion but also a philosophy of life. The separation of religion from a critical reflection on it, i.e. the origin of theology, will come about only much later.

The "crisis of language," notorious in pagan religions and present also in philosophical currents, affected the new religion as well. Just as in the case of the newborn philosophy (see Chaps 1 and 2), it became necessary to create a new "technical" terminology. The mechanism was the same as before: on the one hand, a word or a phrase, taken from everyday language, acquired a new significance in a new conceptual environment; on the other hand, the use of a given phrase in the same circumstances (in a confession of faith, catachesis, liturgy) led to a "petrification" of new meanings. For instance, the "technical" Christian words, such as *salvation* (*salus* in Latin) or *redemption* (*redemptio*), were taken from everyday language, in which the former meant "to be healed" (in the medical sense), and the second "to pay a ransom for a slave" (in the juridical sense).

It is interesting to note that St. Paul, who so condemned all sorts of pagan errors and misconduct, was silent about the Greek image of the world. This seems to testify to the fact that he did not see any major discrepancies between that image and Christian doctrine. Moreover, we can find in the Acts of the Apostles a trace of the fact that Paul's point of view on the cosmos was at least as interesting for Greek intellectuals. When Paul, in Athens, spoke of God who "made the world and all things therein," who "dwelleth not in temples made with hands," and for whom "we live, and move, and have our being" (Acts 17, vv 24–28), the Areopagites listened to him with interest, and only when they heard of the resurrection of the dead did they withdraw their attention.

This story reveals a tension, very typical for Christianity, between its fundamental claim about God's appearance in human history and a philosophical pursuit of the new vision of the world. A taming of this tension will be the principal goal of the newborn Christian theology that had already begun to lay down its own foundations. The new theology had no choice: elements of Greek philosophy had to enter into the construction of its very foundations. Another linguistic revolution turned out to be indispensable: technical terms, already well established in Greek philosophy, had once more to change their meanings to adapt themselves to the needs of Christian theology. This time, the linguistic crisis was even more profound, since religious doctrines of their very nature, because they have a transcendental reference, are stubbornly resistant to all attempts to imprison them by words.

When the author of the Fourth Gospel wrote the Prologue to it, he was undoubtedly referring to Greek philosophical ideas. Even if he did not directly think about Heraclitus or the Stoics or Philo of Alexandria, he certainly took both the concepts and the vocabulary from the store of ideas well known to his contemporaries. The content of the Prologue was not entirely new. *Logos*, the cosmic principle ordering the world, belonged to the Greek heritage. *Logos*, God's Wisdom, creating the world and entering into the human history, was prepared by the Sapiential books of the Old Testament. But exactly at this point a typically Christian shift occurs, causing the above-mentioned tension between the core of the Christian message and its philosophical framework:

"The *logos* was made flesh, and dwelt among us," but "his own received him not" (John 1, vv 11–14).

This tension is a source of all future conflicts between "faith and reason." And conflicts of this kind are irremovable, in the sense that there will always be a lack of proportion between the means of expression and what has to be expressed. Christian faith is authentic only if it incorporates into itself this fundamental gap. Paul knew this very well. The Greeks were looking for wisdom, whereas he preached to them Christ crucified, a stumbling block for Jews and foolishness to the Greeks (1 Cor 1, vv 22–23).

Chapter 7

~

THEOLOGY AND SCIENCE
IN THE EPOCH OF THE CHURCH FATHERS

1. GREEK RATIONALITY AND CHRISTIAN THEOLOGY

*T*he European concept of reason was based on the Greek notion of rationality, but we inherited this notion only after it had undergone substantial transformation by the great adventure of the encounter of Greek philosophy with Christian theology. It was not just an encounter of two strangers. As we have seen in the preceding chapter, Greek philosophy entered into the very foundations of Christian theology and shaped it, as it were, from inside. Of course, there were many external confrontations between Greek thought and Christian doctrine. It is enough to mention the attacks of pagan writers, such as Celsus, Porfirius or Julian the Apostate, and Christian responses such as the *Apology* of Justin. However, a much more important dialogue took place in the thinking of early Christians who read their Bible with heads full of Greek wisdom. Christian theology arose out of this dialogue.

In this early period the word "theology" had a very different meaning (or meanings) from what it has today. We shall not enter here into this question (which is certainly interesting from the historical point of view); instead, in this chapter, when speaking of the theology of early Christian writers, we mean those views which we would today qualify as belonging to theology.

However, we are not interested in the theology of this period in itself, but rather in so far as it has contributed to the "Western spirit of rationality." This is why we want to confront early Christian theology with the science of that period. The question as to which "store of knowledge" in antiquity deserves

41

the name of science is even more complex than in the case of theology. Some of the disciplines that we today call science no doubt existed at that time. These include, first of all, mathematics (mainly geometry), and the three so-called classical sciences: astronomy, acoustics and optics, but also medicine and the beginnings of biology. All these disciplines were very far from our present methodological standards. They were not autonomous from philosophy as they are today. Furthermore, practical knowledge, derived from the sciences, and especially the work of artisans, were seen as science. What we now collectively term "science" had its distant predecessor in a broadly understood philosophy of nature. In the present chapter we are interested in an interaction of early Christian theology with both the natural sciences of antiquity and their philosophical setting.

2. THE CHURCH AND THE ACADEMY: JERUSALEM AND ATHENS

The title of this section alludes to the well-known text of Tertullian:

> *What indeed has Athens to do with the Church? What concord is there between the Academy and the Church? [...] Away with all mottled Christianity of Stoic, Platonic, and dialectic composition! We want no curious disputation after possessing Christ Jesus, no inquisition after enjoying the gospel!*[1]

This text is usually understood as a radical disapproval of pagan wisdom. However, David C. Lindberg questions such an interpretation. Although Tertullian was not an enthusiast of Greek philosophy, he was not here condemning it outright but rather blaming its deviations which could easily lead to heresy. To corroborate this view, Lindberg quotes other passages from Tertullian: "One may no doubt be wise in the things of God, even from one's natural powers. [...] For some things are known even by nature,"[2] or "reason [...] is a thing of God, inasmuch as there is nothing which God the Maker of all has not willed should be handled and understood by reason."[3]

Often quoted is Tertullian's saying: "I believe it because it is absurd" *(Credo quia absurdum)*. This is frequently viewed as a manifestation of his anti-philosophical position. However, Lindberg understands it quite differently. He thinks that "Tertullian was simply making use of a standard Aristotelian argumentative form, maintaining that the more improbable an event, the less likely is anybody to believe, without compelling evidence, that it has occurred ..."[4]

At any rate, the personal views of Tertullian are not that important. The great majority of early Christian writers and the Church Fathers looked with favor on Greek philosophy. Those who did not were an exception, and even they were against the "dangers of wisdom" rather than against wisdom itself. To those exceptional writers belonged, for instance, Tytian, who was skeptical of Greek philosophy, the author of *Didascalia Apostolorum,* who warned his readers against the dangers of pagan literature, and John Chrysostom, who acknowledged the value of pagan schools, but preached that study requires good morals, but good morals do not require study.[5]

A more positive attitude to philosophy was present in the Church from the very beginning, and very quickly grew stronger and stronger. Justin, the first apologist of Christianity, claimed that pagan philosophers had approached the truth because God had given them the rational capacities to do so. Clement of Alexandria went even further by writing that Greek philosophy was indispensable to defend the faith, fight skepticism and develop Christian doctrine.[6]

These were only declarations, but an important process had begun even earlier, namely the process of construing Christian theology based on Greek philosophy. This process was, in a sense, unavoidable. There were two choices: either to be content with a common sense outlook, or to attempt rational reflection and analysis. For people educated in Greek culture, the first possibility was simply resignation rather than a choice. The second possibility meant using the resources of Greek philosophy. The very fact that the first possibility was rejected is not a surprise. What is highly surprising is that the application of Greek wisdom to Christian doctrine had such powerful results.

In fact, the entire Patristic period consisted of a continuous creative process of transforming Greek philosophy into Christian theology. Two great personalities played an important role in this process: Origen and St. Augustine.

Origen (ca 185 – ca 254 CE) was well educated in Greek philosophy. In his doctrine he made ample use of Plato's metaphysics, cosmology and psychology, but he also borrowed some concepts from Aristotle. He was the first thinker to aim to create a synthesis of Christian beliefs with Greek philosophy. He contemplated the world as a harmonious totality and tried to place Christian Revelation within the world image inherited mainly from Plato.

Platonic philosophy fascinated the Church Fathers because it seemed especially "receptive" to Christian ideas. Platonic philosophy, mainly in its Neoplatonic version, also played a key role in liberating Augustine of Hippo from the Manichean sect. His personal experience left a deep mark on his own later teaching. Augustine devoted the rest of his life to reciprocal adaptations of Neoplatonism and Christianity. Contrary to the common opinion that Augustine simply "baptized" Neoplatonism, adaptations had to run in both directions. It is true that Neoplatonism had to be "baptized," but Christian theology, being still "in the making," through its very confrontation with Neoplatonism, plastically adapted its own internal shape to Greek philosophical concepts and to a Greek style of reflection.

Augustine had a great respect for human reason as a precious gift from God. In his letter to Consentius he did not try to hide his emotion when he wrote:

> *Heaven forbid that God should hate in us that by which he made us superior to the other animals! Heaven forbid that we should believe in such a way as not to accept or seek reasons, since we could not even believe if we did not possess rational souls. Therefore, in certain matters pertaining to the doctrine of salvation that we cannot yet grasp by reason, though one day we shall be able to do so, faith must precede reason and purify the heart and make it fit to receive and endure the great light of reason; and this is surely something reasonable.*[7]

The point is not only that the "light of reason" has to pave the way for faith, but also that faith, in our present state, has to prepare our minds to receive full cognition after our death. And it will be a kind of rational cognition. To understand this doctrine better we should not forget Augustine's teaching on illumination: in his view, every intellectual cognition consists of God's special illumination. This is an echo of the Platonic doctrine on the *anamnesis*.

However, it is not only that faith has to precede reason. Further on, in the same letter we read:

> *If, therefore, it is reasonable for faith to precede reason in certain matters of great moment that cannot yet be grasped, surely the very small portion of reason that persuades us of this must precede faith.*[8]

Augustine's strategy, *fides quaerens intellectum*, from the first part of the passage quoted, was completed by a later tradition with the "reverse strategy": *intellectus quaerens fidem*. Both these strategies have their roots in Augustine's teaching. However, we should remember that at that time there did not yet exist a clear distinction between philosophy and theology.

3. THE CHALLENGE OF GREEK COSMOLOGY

A certain world image was obligatory for the educated Greek; it was partially the result of scientific discoveries, but principally it was shaped by the philosophical standards of the time. This image, in its main features, was taken up by the Church Fathers.

For both educated pagans and for Christians of this epoch, the world was finite and had the shape of a sphere, with the Earth situated at its center. With respect to some details of this image there were some clashes between the theological opinions of the Fathers and the views regarded as scientific.

Lactantius belonged to a small group of Christian writers who had difficulties in accepting the sphericity of the Earth. However, his objections were of the "scientific" rather than theological character. He evidently did not know Aristotle's doctrine that "up" and "down" are relative concepts and people on the antipodes walking upside down were beyond his imagination.[9] Lactantius' lack of imagination was mentioned by Copernicus in his Introduction to *De Revolutionibus*, where he wrote:

> *For it is well known that Lactantius, a distinguished writer in other ways but no mathematician, speaks very childishly about the shape of the Earth when he makes fun of those who reported that it has the shape of a globe.*[10]

It was St. Augustine who introduced a theological aspect into the antipode problem. He had no difficulty with walking upside down. He believed, however, that "the inferior part of the earth, which is opposite to where we live" had no inhabitants. And his reason was partly theological: all humanity descends from the one pair, and it is "too absurd to say that some human beings could have arrived from here to there, having navigated the immense space of the ocean."[11]

Some Fathers (St. Augustine included) had a more serious problem with another cosmological question. The biblical account of creation speaks about "the waters which were under the firmament" and "the waters which were above the firmament" (Gen 1, v 7) and such waters were completely unknown to Greek cosmology. Origen interpreted these supra-celestial waters allegorically, comparing them to the "waters of everlasting life" (John 7, v 38), but other Church Fathers tried to supplement Greek cosmology with biblical elements. They claimed that, in contradiction to the Greek view, the outer surface of the world is not smooth but rough in order to hold the waters. St. Ambrose supplied even a "cosmological reason" for the existence of waters above the firmament: they are necessary to cool the axis of the world's rotation in order to save it from burning.[12]

Astrology was not a part of Greek cosmology, but it certainly played some positive role as far as the knowledge of celestial mechanics was concerned. However, making horoscopes was frowned on by the early Fathers. Arnobius, Lactantius, St. Ambrose and St. Jerome, among others, were against astrology, since it could be connected with the worship of the stars. This objection, to some extent, missed the point since, at that time, astrology had already become a kind of secular knowledge.

The case of St. Augustine is especially interesting. The Manicheans, a sect to which the young Augustine belonged, confessed a mixture of Christianity with eastern astral cults. The clever mind of Augustine easily noticed that the astronomical knowledge of the "secular astrologists" ("mathematicians", as they were then called) was much more developed than that of the Manicheans, and this was for him an additional motive to leave the sect. Pedersen comments: "Thus there was an astronomical component in St. Augustine's conversion; perhaps this was not the smallest service done by ancient science to the life of the Church."[13] As time passed, Augustine's appreciation of astrology diminished, and in *De Civitate Dei* it vanished completely. In that work he gave a detailed historical evaluation of it, and acknowledged that it was superstition.

An interesting personage who closes this stage of the interaction between Greek philosophy and science, on the one hand, and Christianity, on the other, is John Philoponus, a Christian and a teacher in the School of Alexandria (6th century BCE). He is often regarded as the last commentator on Aristotle in antiquity. He was certainly a critical commentator; for instance, he denied the Aristotelian thesis of a dichotomy between celestial and terrestrial physics. His arguments in support of this denial were surprisingly correct: he argued that the different colors of various stars testify to their different compositions, and that the fact that stars are composed of something testifies to the possibility of their decay. This criticism evidently had theological consequences: the supra-lunar world is not God's *habitaculum*. One should clearly distinguish between God's absolute transcendence and everything that is created by Him. Philoponus also criticized the principles of Aristotelian dynamics, and in this criticism often referred to experience.[14]

4. CHRISTIAN REVOLUTION

The most revolutionary change that Christianity introduced into Greek cosmology was the doctrine on creation. It was inherited by Christianity from the Old Testament, but the doctrine seemed to be so clear that it was necessary to introduce it only to pagans converted to Christianity (see Acts 17,v 16 seq.). However, in the Prologue to the Fourth Gospel it is not difficult to identify concepts that penetrate more profoundly into the Old Testament doctrine on creation. The Word-*Logos* had already appeared in the teaching of Heraclitus as a force ordering and unifying the World. In the Stoic tradition *Logos* was understood as a principle of the world's rationality (immanent in the world). Philo of Alexandria tried to connect the Greek understanding of *Logos* with the Old Testament usage of the term "word" (*dabar*) with respect to God (e.g. "God said … and it has been made," Gen 1). Independently of the profound theology of the Prologue, the very fact that the Greek concept of *Logos* (albeit Christianized) played a central role in its composition was of great significance for the process of the assimilation of Greek culture into Christianity.

Generally speaking, the Old Testament always understood creation as the absolute dependence of the world on God. Early Christian writers, while not neglecting this meaning, could not resist the temptation to inquire into the "mechanisms" of the world's origin. In the Old Testament there are no traces of this temptation; its appearance at the beginning of Christianity is probably already a symptom of the tendency to approach religious doctrine with heads "contaminated" by the Greek spirit of rationality. Justin the Martyr,[15] Irenaeus[16] and Clement of Alexandria[17] connected the idea of creation with its Platonic understanding as a construction of the world from pre-existing matter. They simply put more emphasis on God's omnipotence, and ascribed creation itself to His free will. But already in the *Shepherd* of Hermes[18] there appears a theological elaboration of the creation concept as a transition from non-being to being.[19]

This doctrine was certainly a reaction against the Gnostic tenet that matter is the "principle of evil." Many Christian writers objected to this view by claiming that even matter is the work of God. This doctrine was developed by Origen[20] who strongly objected to the Platonic teaching on pre-existing matter, and introduced the concept of creation out of nothing (*creatio ex nihilo*) into Christian teaching. However, the full elaboration of the idea of creation belongs to St. Augustine.

The Manicheans ridiculed the biblical story of creation as containing many contradictions and inconsistencies, and Augustine, on leaving the Manichean sect, had to cope with this problem. It is not surprising that he produced several commentaries on the story of creation. The most mature of them is the commentary *De Genesi ad litteram*, completed in 415 CE. In his opinion in the act of creation itself one should distinguish two stages. The first stage was the creation of a shapeless matter *ex nihilo*, the second stage was ordering it into the form of the present universe. One can clearly see here the influence of Platonic cosmology. However, it should be stressed that, in Augustine's view the second stage succeeded the first stage in the logical order rather than in the temporal one. From the temporal point of view, creation was an instantaneous act. Augustine motivated his view by appealing to the Book of Sirach 18, v 1, which in the old Latin translation says: "Qui vivit creavit omnia simul." (He who lives forever has created the sum of things). But, on the other hand, Genesis 1 suggests that the various creatures appeared gradually. How should one interpret this discrepancy?

Here Augustine made use of the Stoic doctrine on *logoi spermaticoi*. Although God created everything at the same instant, not everything was in a ready form from the beginning. Many entities existed only potentially in a seminal form as *rationes seminales* or seed-principles. These entities appeared in their proper form only when suitable conditions arose.

Augustine's *rationes seminales* should not be understood in a biological sense. They were present in the beginning in an analogous way to how old age is present in a young man. Some authors would like to see in Augustine a precursor of the Darwinian theory of evolution. Such a view could be held only when the theory of evolution is understood very broadly. Strictly speaking, ascribing to Augustine evolutionary views would be a historical anachronism: the very idea that some entities could be transformed into other entities was foreign to his thought. His Neoplatonic understanding of ideas in the mind of God would exclude such a possibility.[21]

This "quasi-evolutionary" perspective of Augustine excluded a literal understanding of the six days of creation. In *De Genesi contra Manicheos* he proposed three different interpretations: 1) the six days of creation and the day of rest on the seventh were intended to emphasize the importance of the Sabbath; 2) the seven days denote seven stages in the moral development of man; 3) they denote long epochs in world history.[22] However, in *De Genesi ad litteram – Liber imperfectus*[23] he declared that the story of creation distributed into seven days should be understood as a popular presentation of the development of the world by natural, immanent causes.

The doctrine of creation was so new in comparison with Greek cosmology that it had to raise new questions and make radical modifications, especially with regard to the understanding of space and time. Origen had asked the question: what did God do before the beginning of the world?[24] Augustine supplemented it with another question: where was He before Heaven and Earth came into being?[25] In Augustine's writings one can find many remarks and comments concerning these matters. They culminate in a chapter of *De Civitate Dei*[26] in which he argues that all speculations about space and time before the creation of the world are meaningless since no space and time utterances can refer to God.

These speculations led to the question of the nature of time. Augustine's saying: "What then is time? If nobody asks me, I know it. But if I am asked about it and wish to explain it, I do no longer know it,"[27] became a motif of many dissertations on time. For Origen the biblical expression, "In the beginning God created ...," meant simply that the world began at a certain instant of time, in another instant it will complete its existence and after that moment time will flow indefinitely.[28] This conclusion was queried by Augustine who, relying on Aristotle's physics, claimed that there is no time without motion, and there is no motion before the beginning of the world and after its end. Therefore, he argued, the creation doctrine compels us to accept that God created the world *with* space and *with* time, and that God Himself does not exist in an infinite time and in an infinite space, but exists in eternity, i.e. beyond space and time. As an illustration of this view let us quote a less well-known, but very explicit, passage from *De Civitate Dei*:

> *If we are right in finding the distinction between eternity and time in the fact that without motion and change there is no time, while in eternity there is no change, who can fail to see that there would have been no time, if there had been no creation to bring in movement and change, and that time depends on this motion and change, and is measured by the longer or shorter intervals by which things that cannot happen simultaneously succeed one another? [...] The world was in fact made with time, if at the time of its creation change and motion came into existence.*[29]

There is yet another aspect of the time problem that created interest among some of the Church Fathers. In the Greek tradition time was conceived of as a closed circle. Plato held this view and claimed that circular time is the closest "image of eternity."

The Stoic idea of a cyclic succession of worlds was upheld by Origen, albeit with one important modification. Every subsequent world will be filled in with different events: Moses will not lead Israel out of Egypt, Christ will not be betrayed by Judas.[30] World history is cyclic but time itself is not. In the writings of St. Augustine history ceases to be cyclic, it fully opens. And the reason for this is purely theological: "God forbid, he wrote, that we should ever believe this [the cyclic history] ... Christ once died for our sins and rising again, dies no more ..."[31]

The idea of linear time belongs now to the heritage of our culture. We owe it to the reflection of the Church Fathers on creation and salvation.

5. EVALUATION

David Lindberg asked the question:

> *Did science ... benefit or suffer from the appearance and triumph of Christianity? Did Christianity, with its otherworldliness and its emphasis on biblical authority, stifle interest in nature, as the old stereotype proclaims? Or was there a more ambiguous and subtle relationship?*[32]

One point to be made in answer to these questions is that the Greek sciences started to decline from about 200 BCE, and consequently Christianity cannot be responsible for initiating this process.

In the Patristic period three attitudes dominated: the first was shaped by pagan cosmic religions, with an admixture of Pythagorean, Platonic, Aristotelian and Stoic elements. It contemplated in nature a manifestation of the deity. The second attitude was influenced by Gnostic views. The cosmos was regarded as "the scene of disorder and sin, the product of evil forces." The third attitude was a derivation of Platonic philosophy. It "distinguished clearly between the transcendental world of eternal forms and their imperfect replication in the material cosmos."[33] Christian theology, after a very few initial hesitations, opted for the Platonic attitude. This choice saved Greek science for our culture.

However, we cannot measure the Patristic period by our standards: "... the Church was certainly not calling for the establishment of scientific research institutions nor urging able young men to undertake scientific careers."[34] Nobody did so at that epoch. Doing science was the privilege of a small elite, and only very rarely affected the rest of society.

Nevertheless, let us ask a straightforward question: did the Church Fathers make any contribution to science? In the writings of the Church Fathers there are only occasional references to the classical Greek sciences, and there is no essential contribution to them. Already in the works of the very early Christian writers there appears a "new quality." They treat Scripture as a source of scientific information. St. Augustine advised that in the case of an apparent contradiction between the Bible and a well-established scientific truth, one should give up the literal interpretation of the holy text and acknowledge the priority of reason.[35] However, if such contradictions were absent, Augustine often used a biblical text as a source of scientific information. *De Genesi ad litteram* testifies to this practice. Such use of the Bible to derive scientific knowledge will have deplorable consequences in the future. However, in Augustine's time such a practice was unavoidable. In modern times the natural sciences (geology, biology, cosmology, etc.) stimulate the purification of biblical exegesis from the literal understanding of the world image presented by biblical texts. In the Patristic period there were no developed sciences, and people looked for information about the natural world everywhere, including mythology and the imagination. No wonder the religious authority of the Bible was supplemented by its scientific authority.

On the other hand, we cannot overestimate another process taking place in the Patristic period, namely the process of the assimilation of Greek wisdom, together with the Greek concept of rationality, into European culture. It was not a mechanical process. Greek philosophy was transformed from inside by Christian thought, and Christianity injected into it something of its own vitality. And *vice versa*, Greek wisdom also transformed Christian thought; or, better, it shaped Christianity from the very beginning. It is hard to separate the core of Christian theology from Christianity itself, and Christian theology was formed by both Christian Revelation and Greek intellectual achievements. It is enough to remember that such vital concepts for Christianity as spirit, matter, person, nature and moral law are, to a great extent, of Greek rather than of a biblical origin.

Chapter 8

~

THE MEDIEVAL CONTRIBUTION

1. INTRODUCTION

*A*s the centuries passed, Greek rationality underwent further transformation. Since we are interested in the form that finally gave birth to modern science, we must focus on the corner of the world called Europe. It is there that the mathematical-empirical sciences began their triumphant series of successes. Medieval philosophy was doubtlessly a link between this process and the Greek beginnings. In the present chapter, we try to understand the role this link played in the formation of what could be called "the scientific spirit of rationality."

In contrast with common wisdom, the philosophy of the Middle Ages was neither monolithic nor uniform. To be sure, religious elements played an important role in it, but influences came from various fields: Christian, Jewish and Islamic. They were sometimes constraining, sometimes opening broad vistas. There were many currents and schools, dominant at various times and places. There were many great personalities and independent thinkers not always obedient to higher authorities.

Around 1250 a great shift occurred in medieval philosophy. Before that date it remained under the influence of Platonic, or rather Neoplatonic, thinking. After that date the works of Aristotle, recovered from the Arabs, were already diffused in Europe, and the construction of the great synthesis of Christian theology with Aristotelian science and philosophy was well under way. The shift itself had all the characteristics of a great intellectual revolution; in fact, one of the most profound and far-reaching in the history of human thought.[1] It led to the formation of High Scholasticism, which flourished until ca 1350, but its influence on Catholic thinkers in later centuries cannot be overestimated.

This does not mean that we should underestimate the earlier Platonic, or Neoplatonic, period. It is true that the study of sciences or even theology did not particularly flourish at this time, but we should not forget that to this period we owe the fact that Greek heritage was not lost during the "dark centuries" that separate antiquity and the Middle Ages. Thanks to the enormous efforts of many anonymous people, remnants of the ancient culture that had not been destroyed in the turmoil of the Barbarian wars could survive in cloisters and cathedral schools to await better times. Although the authors working in this period did not elaborate any original doctrine on the relationship between theology and a "secular science," they preserved the doctrine of the Church Fathers on the usefulness of science for theology and on the fundamental concordance between them.

One scholar of the period who greatly contributed to the development of the Scholastic method was Peter Abelard (1079–1142). He thought that universality should be attributed to names, and not to things, and that everything that exists is singular (he thus could be regarded as a precursor of Nominalism). In his work *Sic et Non*[2] he claimed that controversies can often be solved by demonstrating how disputants attributed different meaning to the same words. This idea, codified into the form of detailed rules, became one of the cornerstones of the Scholastic method.

2. HIGH SCHOLASTICISM

St. Anselm of Canterbury (ca 1033–1109) is often regarded as the father of medieval Scholasticism. A similar approach was later developed in Paris and Chartres and soon Paris and Oxford became centers of the new learning. From the close of the twelfth century such schools were organized into associations of students and masters. In this way, "science found its chief institutional home in the universities."[3] The first universities were founded in Paris, Oxford, Bologna and Padua. This process was accompanied by the rediscovery of Aristotle's writings through the intermediary of Arab and Jewish philosophers. The traditional way of thinking was confronted with powerful new ideas. From this encounter medieval Scholasticism took new strength.

Robert Grosseteste (ca 1169–1253) was a pioneer in introducing Aristotelianism in Oxford, although he was still strongly influenced by St. Augustine's Neoplatonism, especially by his teaching on illumination. His views in this respect are called "metaphysics of light." He believed that light is a primordial form of matter, and since the propagation of light is governed by mathematical proportion, mathematics should play a key role in natural science. Similar ideas, with a stronger emphasis on experimentation, were taken up by a Franciscan, Roger Bacon (1214–1292). However, that was not the line that became dominant in the thirteenth century.

Albert the Great (ca 1200–1280) was one of the first to realize that Aristotelian and Arabic science could serve Christian theology. He was interested in almost everything that could be learned by the "natural power of mind." He was an assiduous student of nature, and made many contributions to the Aristotelian heritage of the experimental approach to nature, especially in the field of biology. However, even more important seems to be his influence on the method of philosophical and theological investigation, although in this respect he is considered to be only a forerunner of St. Thomas Aquinas.

St. Thomas (ca 1225–1274), an Italian Dominican, is said to have "Christianized Aristotle." He succeeded in doing so in such a superb way that it finally led to what could be called an "Aristotelianization of Christianity."[4] A cornerstone of the Thomistic synthesis was a distinction between philosophy as the knowledge of everything that can be arrived at by the "powers of natural reasoning," and theology which is the knowledge obtained "in the light of Revelation." He based his metaphysics on the Aristotelian distinction between potentiality and actuality, but enriched it by the distinction between the essence and existence of beings. Then he applied his basic principles to the entire body of philosophical and theological inquiry of the time.

The distinction between philosophy and theology was enforced, in a sense, by the problem of the mutual relations between faith and reason. Latin thinkers, starting from St. Augustine, often gave the primacy to reason, claiming that in the case of conflict "religious sources" should be interpreted in a suitable way. A similar problem arose among Arab thinkers. Extremely influential among them was Averroes (1126–1198), who regarded Aristotle as "the summit of all rational understanding, an infallible guide to knowledge of the world of nature"[5], and was accused of rationalism by the Islamic authorities. It was soon realized that Averroes' teaching could also be in conflict with Christian dogma and indeed, the Latin Averroists (Siger of Brabant and Boethius of Sweden) started to teach theses that were opposed to Christianity. They never abandoned their religious faith, but rather espoused a sort of "double truth" theory.

A reaction was inevitable. There followed several condemnations of which the one in 1227 by Etienne Tempier, the Bishop of Paris, is the best known and had the most significant impact. He condemned 219 propositions believed to be Aristotelian. Among them were several espoused by Thomas Aquinas (they were nullified only in 1325). One of the main goals of this condemnation was to safeguard God's omnipotence. For instance, among the condemned propositions there were statements that the world lasts from eternity, that God cannot create more than one world, and that He cannot move the world along a straight line. "With respect to nature, then, all had to concede that God could do things that were contrary to prevailing scientific opinion about the structure and operations of the cosmos. In short, God could produce actions that were naturally impossible in the Aristotelian worldview."[6]

The impact of this condemnation was certainly weighty but the claim of Pierre Duhem that it marks the birth of modern science seems to be an exaggeration. He argued that the condemnation opened a way to new speculations about the world by weakening the position of Aristotle's physics, and stimulated scientific imagination. However, the condemnation doubtlessly affected theology and indirectly triggered the current of events that finally led to the birth of modern science.

According to an "orthodox Thomistic" thesis, God's omnipotence is limited by many, both logical and metaphysical, constraints. God cannot create things that are self-contradictory, and He is limited by the nature of things. For instance, God may create a horse, or may not create it, but if He decides to create a horse, He must simply implement the abstract idea, or nature, of the horse. And if the natures of all beings are in this way "pre-established," the structure of the world can, in principle, be discovered by pure thinking with only the auxiliary assistance of (everyday) experience. On the other hand, if God's power is absolute, the only way to discover the world structure is to look at the world with wide-open eyes, i.e. by observation and experimentation. From the thirteenth century we can observe a steady shift in theological views from the former position to the latter, and this shift certainly paved the way for the experimental sciences.

This trend was already very clear at the turn of the thirteenth and fourteenth centuries in the Scotistic and Ockhamist movements. Both these currents, although motivated by theological interests, marked a skeptical reaction in philosophy. John Duns Scotus, a Franciscan friar, claimed that the essence of God is His infinity; he emphasized the primacy of His will over His other properties, and consequently His absolute freedom and power. William of Ockham, also a Franciscan friar, taught that God is able to do everything that does not entail a logical contradiction. "The net effect of this teaching was to admit that, in the order of nature, whatever is not self-contradictory is possible; thus, there is no *a priori* necessity in nature's operation, and whatever is the case must be ascertained from experience alone."[7] His principle of parsimony, called nowadays "Ockham's razor" ("beings are not multiplied without necessity") led him to Nominalism and the denial of the existence of "universals;" only names (*nomina*) correspond to general concepts. This had immediate consequences for understanding how we obtain knowledge of the world. In the Aristotelian approach all thought categories had to have their real counterparts and, in consequence, the classification of sciences had to reflect the real "stratification" of the world. Contrary to that doctrine Ockham claimed that the only things that really exist are individuals and their properties.

Ockham's philosophy spread in the fourteenth century and was called *via moderna*.

> *It incorporated a view of the universe that was radically contingent in its being, where the effect of any secondary cause could be dispensed with and immediately replaced by God's direct causality. The theory of knowledge on which it was based was empiricist, and the problems it addressed were mainly those of the philosophy of language.*[8]

3. THE METHOD

Scholasticism is not a philosophical system but a method of philosophizing and learning. This method was often ridiculed as purely verbal and sterile, but unjustly so. Even if its later incarnations were indeed less productive, its original version was a step towards critical thinking.

We should remember that medieval culture (including science, philosophy and theology) laboriously emerged from the ashes of the Greek and Roman world left by wars and the invasions of barbarian tribes. Medieval thinkers started almost from nothing, and had enormous reverence for ancient writers. This is why the medieval culture was "of the overwhelmingly bookish and clerkly character"[9]: "Every writer, if he possibly can, bases himself on an earlier writer, follows an *auctour*, preferably a Latin one. [...] In our own society most knowledge depends, in the last resort, on observation. But the Middle Ages depended predominantly on books."[10]

Our literature often shows a medieval hero as a wanderer and dreamer, but "at his most characteristic [...] he was an organizer, a codifier, a builder of systems. He wanted a place for everything and everything in the right place. Distinction, definition, tabulation were his delight."[11] He formalized war by the art of chivalry, passions by the code of love, and learning by strict rules of leading disputations and analysing texts.

Medieval people were very credulous of books. They inherited a "non-consistent" set of books: Judaic, pagan, Patristic and philosophical of various kinds and provenience. Among them there were chronicles, poems, visions, philosophical treatises, and so on, some in fragments, sometimes in the form of quotations by other authors. Inevitably there were disagreements and contradictions among them. "If, under these conditions, one has also a great reluctance flatly to disbelieve anything in a book, then here there is obviously both an urgent need and a glorious opportunity for sorting out and tidying up."[12] The last sentence is a good description of the Scholastic method.

The principal goal of the Scholastic method was to resolve a contradiction or to answer a question. The first stage was to recognize the state of the question (*status quaestionis*), i.e. to formulate a thesis, establish points of disagreement, and prepare sources by selecting a book by a renowned *auctor*, and other related documents (such as biblical texts, decrees of Church councils, or papal letters). A much discussed text was the *Sentences* (full title: *Libri quatuor sententiarum*) by Peter Lombard, a philosopher of the 12th century. It was a systematic compilation of Scholastic theology. Lecturing on the *Sentences* was a condition for becoming a *magister* in a university.

When the point of disagreement had been established, the argument was made, usually in the form of a disputation between real or imaginary sides. There were quite strict rules for conducting such a discussion. For instance, before answering an objection laid out by the opponent, one had to repeat his point in order to prove that the objection was understood correctly. The chain of objections and answers constituted the core of the dispute. Clear traces of this strategy are found in written works of this period (for example in the *Summa Theologiae* of St. Thomas Aquinas).

The main tools of such a dispute were linguistic and logical analysis. The meanings of words were examined and subdivided and the rules of formal logic applied to find possible errors in the reasoning. No wonder that semantics and logic flourished in medieval schools. These two disciplines were combined into a sort of "practical knowledge", often called dialectics, as opposed to a more mystical approach to theology. The early period of Scholasticism was in fact marked by heated discussion between the defenders of both these trends. In fact, High Scholasticism was born out of this controversy. In this sense Scholasticism can be regarded as a rationalist reaction against the mysticism that distrusted reason as a correct way of doing theology.

4. TRANSMITTING AND TRANSFORMING RATIONALITY

In Chap. 7 we saw that the Church Fathers and early Christian writers preserved Greek culture and the Greek concept of rationality, but it was the Middle Ages that transmitted them to us. However, it was not a passive transmission, but a transmission together with a transformation. The Greek concept of rationality, in particular, had to go through all the abstractions of medieval metaphysics and through all the intricacies of the Scholastic method in order to emerge as the rationality underlying modern science.

The Scholastic method of never using imprecise or ambiguous terms compelled medieval authors to start with careful definitions. But even the most rigorous definition does not guarantee the correctness of conclusions, if the meaning established in a definition is not preserved during the entire process of reasoning. The method of distinctions and subdistinctions of meanings greatly helped the principal goal of avoiding contradictions and protecting the correctness of syllogisms. The evolution of concepts is a key element in the progress of every science. Concepts live in definitions and in the adventures of solving problems, and in these fields medieval thinkers performed a useful service. Modern physics will be born as soon as Scholastic definitions (aimed at grasping the essence of things) change into definitions containing a recipe of how to measure a corresponding property (the so-called operational definitions). It seems that this latter step could not be accomplished without the former preparatory steps.

Not only that. No science can exist without a certain degree of abstraction, and medieval Scholasticism was certainly a very good exercise for philosophers and theologians in performing various kinds of abstraction. It is true that the method of abstraction was always associated with doing philosophy, but in this respect medieval philosophy was very special. Abstraction became an art, subject to rigorous rules of Scholastic procedures and logical schemes. When these procedures and schemes change into mathematical patterns, we shall already be within the method of modern science. Of course, this change could not be instantaneous, but the process of maturation is a link between the Middle Ages and Modernity.

When we are thinking about the medieval contribution to the modern concept of rationality, we should not forget one important element, the idea of God as a supreme guarantee of the rationality of the world and our rationality. In Chap. 1.3, when we spoke about the erosion of mythical religions under the impact of the new-born philosophical thinking, we mentioned that in some philosophical systems there appeared a god or a deity that was not a subject of worship, but rather a sort of "closure" of a given system (e.g. Aristotle's First Mover). Such a god or deity, being himself a conclusion of premises constituting the system, could not guarantee the system's rationality. Medieval metaphysics and theology radically changed this state of affairs. The god that was indispensable as a "closure" of the metaphysical system was at the same time the Christian God to be worshipped and loved[13]. Never before had rationality been given such a powerful foundation.

It was not only the Scholastic method that contributed to the formation of the scientific method: the content of many medieval disputations was also a link in the chain of events leading to modern science.[14] A good example is the concept of laws of nature, which evolved from Scholastic disputes on God's omnipotence. We saw in Sect. 2 how, as time passed, a tendency shifted from ascribing to God power limited by some metaphysical constraints, e.g. by the nature of things, to the idea of His unlimited or absolute power dependent only on His own will. God's constraints clearly transfer into the constraints of the behavior of nature. If there are immutable natures of things, which even God must respect, we can hope to discover the functioning of the world by pure speculation. If there are no such constraints, the only way to discover the functioning of the world is by observation and experimentation. The medieval world, for this lack of necessity, was a *contingent* world. Of course, even the contingent world does not behave in an irregular and unpredictable way. Already the Church Fathers introduced the distinction between God's absolute power (*potestas absoluta*) and His ordained power (*potestas ordinata*). "The former considers God's power as such, and recognizes no limits to it, no confining law or order, except for the principle of noncontradiction; the latter considers God's power inasmuch as it is actualized or realizable in an order of things."[15] Medieval thinkers developed this distinction and made out of it a subject of heated disputes. The regularities in the behavior of the world are but vestiges of God's ordained power. By discovering them we can learn something about God Himself and His work. To study this became the task of Natural Theology that flourished in England at the time of Newton. The term itself, "laws of nature," was also of "Scholastic, and perhaps even older, origin."[16] In early modern science it was connected with the idea of „universality and necessity" found in nature. And in Leibniz's writings the distinction between God's absolute and ordained power became the distinction between logical necessity and physical necessity, so important in modern thinking about the world.

PART III

~

DISCOVERY OF THE METHOD

When we look at the world one of the first questions to strike us is to do with motion: why do things move? Do they really move? Motion is the sum of rests. How, by adding zeros together, can we obtain anything different from zero? And so on.

Such questions, asked by the Greeks, created kinematics, the science of motion. Zeno of Elea, with his antinomies of motion (Can the flying arrow reach its target? Can Achilles overcome the tortoise? and others), went even further. In his persistent questioning we can identify the seeds of serious problems: the problem of continuity and infinite divisibility, the nature of infinite sets, the problem of limit and of instantaneous velocity. Twenty-five centuries and new branches of mathematics (set theory, topology and calculus) were necessary to resolve Zeno's paradoxes.

Are they indeed resolved? Mathematics is a purely formal science and as such it does not refer to reality, but motion belongs to the real world, and consequently no problem of motion can be solved by mathematics. Correct. But mathematics is used to model the world, and if this is the case, it becomes physics. First, we must correctly guess a mathematical structure, and then interpret it as representing the structure of a certain aspect of the world. The history of science teaches us that more often than not our guesses and our interpretations are correct. And when this is the case, the mathematical model of a physical phenomenon that we have created rewards us with empirical predictions. If they agree with actual experimental results, we can confidently say that our model works well. In this sense, calculus, together with other coworking mathematical disciplines and suitable physical interpretations, solve the problem of kinematics. This is the subject matter of Chap. 9.

But then comes the problem of dynamics, i.e. the problem of motion under the action of forces. The laws of dynamics (also called the laws of motion) must be formulated. The first such laws were formulated by Aristotle. It is even possible, by studying his texts, to reconstruct his two laws of dynamics. The first says that a body, acted upon by no force, remains in the state of absolute rest, and the second states that a force is necessary to move the body, i.e. to change its state of rest into motion. The first law, just as in Newtonian mechanics, establishes the "standard of motion" (in this case the state of absolute rest); the second determines the action of a force as causing a deviation from this standard. Aristotle's laws were reconstructed in close analogy to Newton's laws and by ascribing to Aristotle some Newtonian concepts, but, in fact, there is a long evolution of concepts that separates Aristotle from Newton. This is the principal reason why our reconstruction of Aristotle's dynamics is only a strong stylization of his rather fuzzy ideas. We tell this story in Chap. 10.

To follow the evolution of the concepts that finally led to the correct formulation of the laws of motion would require several thick volumes. We have therefore chosen to report, in Chap. 11, one of the last episodes in this long chain of events. What started with Zeno's flying arrow approached its end with the analysis of gunfire. In the 16th century Nicolo Tartaglia studied the motion of cannonballs. Because Aristotelian physics forbad mixing natural and enforced motions, the accepted theory was that a cannonball moves first along a straight line and, when its impetus is exhausted, falls directly down towards the Earth's center. Tartaglia, after some hesitation, came to the conclusion that the trajectory of a "projectile" is curved along all its length. The Aristotelian doctrine had been refuted, although the precise shape of the trajectory remained unknown.

The problem of the trajectory's shape can be eliminated by considering free-falling bodies, but the Aristotelian doctrine created a stumbling block here as well. According to Aristotle, the velocity of a falling body depends on its weight. Gianbattista Benedetti, a disciple of Tartaglia, considered this problem. Although at the very beginning of his early work Benedetti declares that he will destroy Aristotle's theory, the error was too deeply rooted to be able to eradicate it at the first approach. Benedetti argued that only bodies "of the same nature" fall with the same velocity regardless of their weights. The correct answer had to wait until Galileo Galilei, the disciple of Benedetti, came on the scene.

It was Galileo who developed the correct theory of a material point moving under the action of a constant force, i.e. the theory of uniform motion and that of uniformly accelerated motion of a material point. Although Galileo never used the term "principle of inertia," the fact that he applied this principle to his theory of uniform motions makes him its discoverer. The theory of free-fall is also of the greatest importance because it provides a "clinical case" of a uniformly accelerated motion that can be isolated from the "rest of the universe" and studied independently. These two special cases are the cornerstones of Galilean mechanics.

Isaac Newton, as he himself remarked, was able to see further than others because he was standing on the shoulders of giants. The giants, like Copernicus, Kepler and Galileo, did invaluable work, but it was Newton who not only laid the foundations but also constructed the edifice. In Chap. 12 we take a closer look at Newton's Principia, not to do the work of an historian of science, but rather to grasp the main features of the new mathematical-empirical method which from now on will be the only method admissible in physics. We do this by focusing on the Newtonian laws of motion. The problem has finally matured to the point of being solved: concepts referring to motion correctly defined, and the mathematical structure to model motion (calculus) ready to do its work.

The results obtained with the help of the new method quickly accumulated, and soon produced a new image of the world. It will last and be mandatory for almost three centuries, but the most permanent result of the story told by us in the preceding chapters is the method itself. It has created a new way of understanding, most probably the only authentic and viable way of understanding the world.

At the end of Part One we proposed a hypothesis that the world is rational, in the sense that we should ascribe to it a property owing to which it can be rationally investigated. There are many methods with the help of which we could try to understand the world, but only when people started using the mathematical-empirical method did progress in understanding the world become so rapid that it had no parallel in any other field of human activity. This allows us to formulate our hypothesis more precisely: we should ascribe to the world a property owing to which it can be investigated with the help of the mathematical-empirical method. In this sense, one often says that the world is mathematical. In Chap. 13 we argue that the question why the world is mathematical is not a trivial question. It is perhaps the most important philosophical question raised by the very existence of modern science.

Finally, in Chap. 14, we take a look at how the mathematical method works in modern science. The miracles of mathematics were not exhausted in creating general relativity, quantum physics and chaos theory. Its power is still at work. It drives us to the final goal, the full unification of physics and the ultimate understanding of the world. The inevitable question arises: does the mathematical method of investigating the world have any limits? We can only suspect that if such limits do exist, it is the mathematical method itself that would be powerful enough to discover them. And this indeed seems to be the case.

Chapter 9

∼

ACHILLES AND THE ARROW

1. THE DIALECTIC OF MOTION

*T*he eye focused on the target. Attention increased to its maximum. Decision. Now! The arrow pierces the air, a disappearing dot flying to its goal.

This beautiful gesture of a sportsman aiming at the target (not to mention the less beautiful art of killing) became the subject-matter of a philosophical analysis at the very beginning of European thinking.

> *... if everything when it occupies an equal space is at rest, and if that which is in locomotion is always occupying such a space at any moment, the flying arrow is therefore motionless.*

This text was written by Aristotle[1] who presented in it one of the outspoken antinomies coined by Zeno of Elea against the possibility of motion. The problem had the following historical framework.

The phenomenon of motion, the transition from one place to another or, more generally, any kind of change, is so ubiquitous that one can either see it everywhere, or overlook it entirely. The former possibility was adopted by Heraclitus of Ephesus (6th to 5th centuries BCE). Only a few fragments of his writings have been preserved. Everybody remembers his sayings: "Everything flows" or "One cannot enter twice into the same river." Let us add one less well-known one: "We are afraid of death, but we have, in fact, died many times."

Heraclitus wanted to tell us that the process of motion or of any change involves a certain contradiction: something is in a given state, and at the same time it leaves this state. Something like dying and being born simultaneously.

A drastically different perspective was adopted by Parmenides of Elea (5th century BCE). Some philosophers claim that he was the greatest discoverer in philosophy since it was he who discovered the concept of being. It is the most general concept one can imagine. Everything that exists (in any sense of the word) is a being. Parmenides' discovery is encapsulated in his adage: "Being exists, and non-being does not exist." We do not want to enter into metaphysical disputes, but there is no doubt that his dialectic introduced a lot of misunderstanding into the problem of motion. If it had a positive effect on the problem at all, it was to encourage people to refute his various arguments against the possibility of change.

The starting point had the character of a verbal trap: the only change the being can undergo could be the change from being to non-being; but non-being does not exist; therefore being cannot change. One of Parmenides' disciples, Zeno of Elea, was able to go beyond this purely verbal game, and to identify the seeds of real problems in this dialectic. His famous antinomies of motion (the above-quoted antinomy of the flying arrow is one of them) were intended to be mathematical propositions of a sort showing the contradictory character of motion. These contradictory properties had not only fascinated ancient thinkers, but also turned out to be the first links in a long chain of future developments.

Two further antinomies are the one called "bisection," and the other "Achilles and tortoise." The former asserts the impossibility of motion, because "that which is in locomotion must arrive at the half-way stage before it arrives at the goal." And then the half-way of this half-way, and so forth, to infinity. The latter is but a more dramatized version of the former: Achilles cannot overtake the tortoise, "since the pursuer must first reach the point whence the pursued started, so that the slower must always hold a lead."[2]

In contrast to the sterile reasoning of Parmenides, the paradoxes formulated by Zeno point to real problems. From our present perspective we can identify the following: the problem of continuity and infinite divisibility, the nature of infinite sets, the problem of instantaneous velocity and, consequently, of the limits of infinite sequences or functions. Twenty-five centuries were necessary to correctly formulate all these problems, and such chapters of mathematics as calculus, set theory and topology, turned out to be indispensable to solve them. The evolution of concepts in science is a laborious process.

2. ACHILLES AND THE TORTOISE
AFTER TWENTY-FIVE CENTURIES

There is a saying that the history of past events changes depending on our present knowledge. This saying is even truer with respect to the history of science than with respect to other histories. The great problem of motion had its beginning in Zeno's paradoxes, but to correctly assess this starting point we must look at it from our present point of view.

Every freshman in physics knows very well that no problem referring to motion can even be approached without a sufficient command of calculus. The suspicion arises that all Zeno's troubles with motion were due to the fact that he lacked this powerful tool. This is the opinion of many physicists and mathematicians. For instance, Carl Boyer says that the concept of derivative solves all paradoxes, and it is only the weakness of our imagination that is responsible for our difficulties in grasping the idea of continuity and limit.[3] Should we then consider the problem as settled? The long list of serious authors (Pierce, James, Russell, Whitehead, Bergson, Whitrow, etc.) who quite recently had troubles with Zeno's paradoxes seems to falsify this supposition. Whitrow thinks that calculus alone cannot solve these paradoxes since the time problem is involved in at least some of them, and the problem of time remains beyond the reach of pure mathematics. In his opinion, logical antinomies appear whenever we connect the mathematical concept of continuity with the idea of transience, which is a non-mathematical idea.[4]

Henri Bergson (1859–1941) said once that any interpretation of his philosophy that does not put the idea of transience at its very center, would inevitably be a misunderstanding. He claims that the empirical sciences are unable to capture what is at the core of reality, namely the idea of "flowing" and continuity. Physics – he argues – eliminates true motion and the true aspect of temporality from the image of the world by changing "that which flows" into "that which is spatial." Bergson calls this spatialization or geometrization strategy, and claims that it falsifies reality.

By adopting such an approach, Bergson "has solved" Zeno's antinomies.[5] Strictly speaking, in his view, there is nothing that should be solved: the problem is badly posed from the very beginning. Zeno's error consisted in his attempt to conceptually analyze what can only be grasped by intuition. The spatialization strategy is here at work – the covered distance replaces motion, and in this replacement the idea of transience is lost. If we eliminate our intuition, the arrow will not reach its target, and Achilles will never overtake the tortoise.

In Bergson's doctrine we can hear an echo of Aristotle's view that mathematics cannot adequately deal with the full diversity of the world and the richness of human experience.

3. THE MIRACLE OF THE METHOD

The views typically represented by Bergson are based on a very common misunderstanding. It consists in believing that physical theories should *describe* a certain fragment of reality, i.e. that they should copy it (as faithfully as possible) in some linguistic material. Consequently, physics, although it deals quite well with a "quantitative aspect of motion," is accused of being unable to put into formulae the intuition of flowing and transience. The point is, however, that the aim of physics is not *to describe* some domains of reality, but rather *to model* them. We do not demand from a physical model that it should be a "smaller copy" or a "translation into formulae" of the aspect of the world that is to be modeled. In the method of modeling we assume that *a certain mathematical structure represents a certain aspect of the world.* Let us explain the meaning of this statement.

First, we must have a mathematical structure. Very often it consists of an equation (or of a set of equations), but always together with all necessary conditions indispensable to give a correct meaning to the equation (such as the space on which the equation is defined, the initial or boundary conditions necessary to solve it, etc.). History teaches us that there are various ways of arriving at such a structure. It can be a flash of intuition, the method of trial and error, or laborious thinking. In any case, it must be founded on a deep knowledge (both theoretical and experimental) of a given problem, and usually it is carefully prepared by gradually building on the work of predecessors.

The mathematical structure, when correctly guessed or identified in some other way, is not treated as a part of pure mathematics, but rather referred to the world as a part or an aspect of its structure. This is the most subtle part of the whole procedure. Roughly speaking, by a structure we understand a network of relations between some elements (often containing also relations between relations), the nature of which remains irrelevant. If such a structure is encoded into mathematical symbols, it is said to be a mathematical structure. And if such a mathematical structure is referred to the world, it is tacitly assumed that the same network of relations constitutes *the structure of the world.*

This assumption is a kind of decree, on the strength of which we assign to the world (or to some of its aspects) the same structural properties that form a given mathematical structure. In this sense, the explanation of the world in physics is always a *structuralist* explanation. Non-structuralist properties of the world (such as the nature of elements between which the relations hold) are not taken into account in the modeling process.

Any structure is an entirety. One speaks of aspects of a structure rather than of its parts. In the modeling process an *idealization* is always involved. However, it does not consist in assuming that a given mathematical structure only approximately represents the structure of the world, but rather that it represents only a certain aspect of the world's structure and ignores other aspects. However, in many cases it is assumed that it does this precisely. This can be clearly seen when we are modeling that aspect of the world that is not subject to direct observation by our senses. For instance, it is meaningless to say that the wave function of an electron approximates its quantum state, because everything we know about the quantum state of the electron is through the mathematical structure of our model, in this case, the wave function together with the theory of Hilbert spaces which is necessary for the wave function to be correctly defined.

So far we have said nothing about the experimental side of the physical method. However, this side is vital. Experiments appear at two ends, as it were, of the method. First, in the preliminary phase. One could hardly imagine choosing the correct mathematical structure without prior empirical investigation. The final decision as to which mathematical structure should be adopted, might be a sort of illumination, but such an illumination does not jump out of nothing. It must result from a deep knowledge including also knowledge of the empirical side of the problem. Second, empirical investigations appear in the last phase as a final justification of the entire process. If, by manipulating a given mathematical structure (interpreted as an aspect of the world's structure), we are able to deduce from it empirical predictions, and if these predictions turn out to be in agreement (within measurement error) with the results of actual experiments, then we are entitled to claim that our model does indeed represent the modeled aspect of the world. It is a miracle that this method works. And it works very well!

In spite of the fact that empirical investigations appear "at the ends" of the method, they belong to its very core. They constitute the only justification of our, otherwise purely conventional, decision to make a bridge between mathematics and the world.

4. ANTINOMIES OF TRANSIENCE

Let us go back to Zeno's antinomies. Who are then right, those who, with Boyer, claim that calculus has liquidated them, or those who, with Bergson, assert that the empirical sciences will never be able to grasp the flow of motion and the transience of time? As usual in such situations, the correct answer is somewhere in between.

Motion and time are doubtlessly aspects of the world's structure and, of course, as such they are beyond the reach of pure mathematics. This is the point for Bergson. But, on the other hand, Zeno's antinomies cannot be solved without the help of mathematics. In this respect Boyer is right. However, he overlooks the fact that, besides mathematics, one must also employ the methodological procedure of modeling the world. Preciseness of mathematical structures, combined with their function of representing some aspects of the world structure (in this case, motion and time), entirely liquidate the antinomies.

In modern physics, it is calculus that provides the structure necessary to model motion and time. The key concept of this structure is the concept of a real function of a real variable. Independent variables of such a function run through the line of real numbers (or through some of its intervals), and the function also assumes its values in real numbers (dependent variables). If we construct a model of motion, the independent variable (usually denoted by t) represents time, and the dependent variable (often denoted by s) represents the distance covered in time t. We thus have a functional dependence of distance s on time t, and the derivative of distance with respect to time, ds/dt, represents the instantaneous velocity.

In this model, time instances are thus represented by real numbers. The essential thing is that these numbers form a part of a bigger mathematical structure, namely of the real line R, and all structural aspects of the real line are ascribed to time. Owing to this fact, (almost) all effects connected with our experience of time are present in this model: the sequence of moments, continuity of this sequence, divisibility of any time interval into arbitrarily smaller intervals, etc. The only element that is lacking is the psychological impression of transience or flow. It is not a part of the model, because it constitutes the subject-matter of psychology and not of physics. However, everybody who knows something about the natural order on the real line and its topology, immediately sees that the precision provided by the model infinitely surpasses the precision of our psychological experience, not to mention the preciseness of any psychological analysis. For instance, the length of our experienced "now" is up to 0.6 seconds,[6] whereas in the considered model "now" can be modeled by a single point.

However, in this model there is nothing that would correspond to the irreversibility of time. This is perhaps the most painful element of our personal experience. Two remarks are to be made. First, the correspondence between mathematical structures and some aspects of the world structure are more often than not surprisingly accurate, and if in a good mathematical model some element does not appear (which we would otherwise have expected), this does not usually happen without reason. It cannot be excluded that the irreversibility of time is not its essential property. For instance, in contemporary quantum field theories there are strong reasons to believe that the irreversibility of time is only its macroscopic property, and that on the microscopic scale the direction of time can be reversed (e.g. antiparticles can be regarded as particles living in the reversed time). Second, it is not true that the irreversibility of time cannot be modeled mathematically. To this end, we must simply use other models of time, based on different mathematical structures: for instance, the model which is constructed in statistical thermodynamics.

Let us go back to the time model discussed above. It is a part of a larger model of motion. As we have seen, it very effectively models the process of change as a function of a given changing magnitude of time. The derivative of this function with respect to time represents the instantaneous velocity of this change. Although a psychological feeling of "flowing motion" is not a part of this model, Bergson is wrong when he claims that physics "freezes" the true motion, and "spatializes" the true time. The concept of derivative very precisely models the process of change, its continuity and its instantaneous velocity. Let us emphasize this once more: our "direct intuition," so frequently alluded to by Bergson, cannot compete with the mathematical model discussed as far as the preciseness of the magnitudes involved is concerned.

Of course, an elementary knowledge of calculus is not enough to fully resolve Zeno's paradoxes. As is well known, to correctly define such notions as continuity, limit of a function and derivative, advanced tools of set theory and topology are needed. And, as we have tried to explain, to go from pure mathematics to physics one should employ the method of mathematical modeling. It is only precise mathematics together with the physical method of modeling that smooths out Zeno's paradoxes.[7]

One more comment. Idealization, which is an unavoidable element of the scientific method, does not falsify the reality given to us by our cognitive intuition (as was claimed by Bergson and his followers); it is a methodological strategy, owing to which we are able to face the enormous (perhaps infinite) richness of reality. To see this, it is enough to turn to our investigations of the subatomic world. Mathematical models of quantum objects do not falsify them since they are not evident either to our intuition or to our senses. Without mathematical models we could only tell stories about "invisible components of matter."

5. THE EVOLUTION OF PROBLEMS

The history of science is first of all the history of solving problems. It is sometimes said that a problem cannot be solved if it is incorrectly posed. But, in fact, just the opposite is true: no problem can be correctly posed before it has been solved. Only from the perspective of the known solution are we able to see all the nuances and conceptual traps. This is why at the beginning of an evolutionary chain problems are formulated dimly, but only those that are formulated creatively initiate processes that lead to future successes. Other problems are mercilessly eliminated by the history of science. From the point of view of physics, Parmenides formulated the problem of motion in a non-creative manner, and his approach was simply ignored by physics. On the other hand, Zeno formulated his paradoxes in an ambiguous way, but in his formulation there was a creative element. As we have seen, they finally led to the concepts of function, its limit, derivative, set theory, topology and many other concepts of fundamental significance for mathematics. This was one evolutionary branch of the mathematization of motion process. The second branch was of a more physical character. Dynamics is the science of motion under the influence of forces. A creative formulation of the fundamental dynamical problem was very difficult and highly complicated. The only way out of such a situation was to reduce a difficult problem to an easier one, and try to solve it. If we are unable to answer the question of what happens to a body acted upon by many forces, let us try to answer what happens to a body acted upon by no forces. This question leads to the First Law of Dynamics.

The problem of motion will not be fully solved until these two evolutionary branches, the one leading to calculus and the other to the laws of dynamics, meet together. This will happen in Newton's work.

Chapter 10

~

THE DYNAMICS OF ARISTOTLE

1. PHILOSOPHICAL BACKGROUND

*T*he effects of world wars usually last a few decades, rarely a century, but words written by a great thinker can remain highly influential for many centuries. Very few written pages had a greater effect on the science of motion than Aristotle's *Physics*. It was only in the 17th century that Galileo and Newton corrected Aristotle's errors, but even these successes were, in a sense, caused by Aristotle as a reaction against his doctrine. We must thus take a closer look at Aristotelian dynamics.

Let us first recall the scenery. On the one side, there is the world of Heraclitus, flowing as a river, into which one cannot immerse oneself twice. On the other side, the static world of Parmenides from which every change is excluded. In spite of all these speculations, everyday experience, supported by common sense, tells us that motion does exist, and that there is in it a certain continuity, owing to which the moving thing preserves its identity. How is it possible?

It was Aristotle who tried to answer this question, and it was his theory of *actuality* and *potentiality* that was supposed to solve the riddle. For a change to be possible a body must be in the state of potentiality to accept something which it is still lacking. The change is completed if this something is actualized. "The fulfillment of what exists potentially," writes Aristotle, "in so far as it exists potentially, is motion."[1] In Aristotle's view, this conception, on the one hand, guarantees the reality of change and, on the other, allows the changing thing to preserve its identity. Even today there are philosophers who highly value Aristotle's solution but, in our opinion, it would be hard to defend it against the accusation of purely verbal analysis. We could agree that the idea of potentiality and actuality created a handy terminology in speaking about change and motion, but from language to reality there is a long road.

After giving his definition, Aristotle writes:

> *Examples will elucidate this definition of motion. When a buildable, in so far as it is just that, is fully real, it is being built, and this is building. Similarly, learning, doctoring, rolling, leaping, ripening, aging.*[2]

We can see from these examples how broadly Aristotle understood motion. He was a biologist rather than a physicist (using our present standards), and that is why he was thinking in terms of organic changes rather than in terms of physical motion. From among his examples we would today accept as motions perhaps only rolling and leaping, and even they would not be the best instances to begin a kinematical analysis. However, such an approach was in agreement with Aristotle's philosophical standpoint. As we remember, he believed that science should face the richness and complexity of the world in its totality, and that the best methods for doing so are causal explanation and qualitative analysis.

2. TWO LAWS OF ARISTOTELIAN DYNAMICS

This does not mean, however, that Aristotle did not try to deal with the quantitative aspect of motion. Basing oneself on his various texts, mainly on the last parts of his *Physics*, it is even possible to reconstruct his laws of dynamics. Obviously, his dynamics strongly depends on his philosophy, and we should not forget that at that time science and philosophy constituted one body of knowledge.

As is well known, Aristotle distinguished natural and enforced motions. In his view, every being tends to its goal, and this tendency is the principal cause of all motions. For a body subject to a local motion (i.e. changing only its place), the goal is to reach its natural place. The center of the Earth is the natural place for heavy bodies, whereas the circumference of the world is the natural place for light bodies (such as fire). A body moves naturally if it is moving to its natural place. Such a body can deviate from its natural motion only under the action of a force. If such a force is acting, the motion is an enforced motion.

The reconstruction of the Aristotelian "equations of motion" is not unique. A certain freedom results, first of all, from the fact that Aristotle did not have at his disposal sufficiently precise concepts referring to motion. He used a language based on intuition. For instance, he often employed the term "body" where we would use "mass," or "acting factor" where we would say "force." Moreover, Aristotle tried not to use idealizations; he tried to face the world in its full richness and complexity. But when attempting to write a dynamical equation, he simply had to make idealizations, and he did them implicitly, with no control over factors that had been neglected. And this, of course, effected the result.

Making use of this freedom in interpreting Aristotle's texts, let us formulate his "Laws of Dynamics" in as close analogy with Newton's laws as possible.

The First Law of Dynamics: If a body is acted upon by no force, it remains in the state of (absolute) rest.

The Second Law of Dynamics: The force F acting on a body of mass m is proportional to the mass m and velocity v which the force F imparts to the body. This law can be written in the form of the equation $F = mv$.

Let us notice that in formulating the above laws we have used our present terminology and our present symbols.[3] Both these laws are compatible with each other, in the sense that if $F = 0$ then from the Second Law it follows that $v = 0$, i.e. the body is at rest (which is asserted by the First Law).

It should be mentioned that another interpretation of Aristotle's physics is also possible. If we take into account the resistance of the medium, his Second Law does not correspond to Newton's Second Law, but rather to the Stokes Law which asserts that the force of resistance acting on a body moving in a viscous fluid is proportional to the velocity which the force imparts to the body. An argument on behalf of this interpretation could be that Aristotle believed that a vacuum could not exist, and consequently all bodies move in a resistant medium. If, in such conditions, no force is acting, the body indeed remains in rest.[4]

Both these interpretations of Aristotelian physics are, in a sense, historical anachronisms, and both of them depend on how we translate Aristotle's intuitions into our present concepts. However, such "anachronisms" seem to be justified by the view that the logic of scientific evolution can be properly evaluated only from the perspective of the solution achieved.

3. THE PRINCIPLE OF INERTIA

The First Law of Dynamics in Aristotle's formulation says: "everything that is in motion must be moved by something."[5] In spite of his objections against idealizations this formulation is, in fact, based on a quite strong idealization. If a horse ceases pulling a carriage, the carriage stops, but if I throw a stone, it continues moving in spite of the fact that my hand ceased to act on it. Similarly, Newton based his First Law on an idealization: we never observe a body moving indefinitely in a strictly uniform way if no forces are acting on it. In Newtonian mechanics we explain the fact that the body finally stops by recourse to friction and the resistance of the medium. However, Aristotle had to look for some "causes" that would explain why the stone once thrown continues to move. He did that in a rather unclear text:

> ... in point of fact things that are thrown move though that which gave them
> impulse is not touching them, either by reason of mutual replacement, as
> some maintain, or because the air that has been pushed pushes them with
> a movement quicker than the natural locomotion of the projectile wherewith
> it moves to its proper place.[6]

Later commentators understood this text in the following way. A thrown object, for instance a flying arrow, pierces the air, pushing it forward. The increased pressure in front of the arrow causes the air to move to the back of the arrow which, in turn, pushes the arrow forward. It would not be worthwhile to mention this pseudo-explanation if it were not for the fact that it attracted the attention of thinkers and, in this way, was a major factor in future developments.

It is interesting to note that in Aristotle's *Physics* there is a text that seems to anticipate Newton's formulation of the First Law:

> ... a thing will either be at rest or must be moved ad infinitum, unless some-
> thing more powerful gets in its way.[7]

Unfortunately, this was intended by Aristotle as an argument, by *reductio ad absurdum,* against the possibility of an "infinite motion." If it were not for Aristotle's unwillingness to make idealizations, we would perhaps have had the laws of dynamics many centuries earlier.

4. DYNAMICAL STANDARDS

As we have seen above, the First Law of Dynamics is the consequence of the Second Law (both in Aristotelian and Newtonian dynamics). Is it then necessary to assume it as a separate law? The point is that the First Law is something more than a mere consequence of the Second Law. In fact, the First Law establishes the "conceptual perspective" of the entire mechanics. If we want to know the effects produced by the action of a force on a body, we must have at our disposal a "dynamical standard" with respect to which these effects could be estimated. Such a standard is given by a situation in which no forces are acting. In other words, we must know how the body behaves when there are no forces. If this behavior deviates from the standard, we must look for some dynamical cause of such a behavior; in other words, we must assume that a force is acting. As we can see, without the "dynamical standard," the concept of force is meaningless. To use Aristotelian language, we could say that the "dynamical standard" defines the "natural" state of a given body. As long as a body is in its "natural state," no dynamical justification is required; it is a deviation from the "natural state" that must be justified (however, we should never forget that dynamics is an empirical theory, and it is the agreement with experiments that finally justifies the entire system).

We see, therefore, that the First Law of Dynamics, as establishing the "dynamical standard," is logically indispensable. In Aristotle's dynamics it is absolute rest that is the standard; in Newton's dynamics it is uniform motion.

Aristotle's physics had support in his cosmology. Since the center of the Earth is situated at the unmoved center of the universe it seemed natural to identify the dynamical standard with the "rest at the Earth's center." Everything that moves with respect to the Earth's center moves naturally. All other motions are "enforced"; a force that causes them must exist.

The Aristotelian world view became less and less popular from the time of the Copernican revolution. Finally the "standard of absolute rest" ceased to exist. This certainly helped Galileo, Kepler and Newton to make the final step. If there is no natural rest, the next "natural candidate" to become the dynamical standard is uniform motion.

Of course, the reality was much less logical than our present analysis would suggest. There were many conceptual traps and misunderstandings before a relatively clear panorama finally appeared. And with all the consistency of the new system there was still a gap in the Newtonian image of the world. The dynamical standard of uniform motion was slowly establishing a new scientific paradigm, and it led to surprisingly correct empirical predictions, but it had no firm cosmological support. The "post-Newtonian" cosmology was a conglomerate of Copernicus' and Kepler's achievements, the remnants of the Aristotelian heritage, Descartes' speculations, a particular interpretation of Euclidean geometry, together with Newton's theory of universal gravity. Although, starting from the work of Galileo, it became more and more clear that there is only one physics governing both "heaven and Earth," the full unification of these two physical domains was still a postulate rather than a firmly established scientific fact. Yet in the 19th century some attempts to construct Newtonian cosmology led to difficulties and paradoxes. It was only Einstein who in his theory of relativity drew final conclusions from the existence of the dynamical standard of uniform motion, which, together with the postulate that the speed of light is the maximal physical velocity, became a decisive step towards a cosmology that was fully integrated with the rest of physics. But that is another story …

Chapter 11

~

THREE GENERATIONS:
FROM TARTAGLIA TO GALILEO

1. STROKE OVER S

*O*n reading the last chapter, anyone who has had any contact with classical mechanics will have noticed that there is only a slight difference between Aristotle's Second Law of Dynamics and that of Newton. In Aristotle's version, force is proportional to mass and velocity, whereas in Newton's version it is proportional to mass and acceleration. If, in agreement with the present notation, we denote velocity by s' (derivative of distance with respect to time), and acceleration by s'' (second derivative of distance with respect to time), then this difference is reduced to only one stroke over s. But, in fact, the difference is enormous. Between Aristotle and Newton a long evolutionary chain extends. To add this little stroke over s, new concepts had to be elaborated, many problems posed, some of them solved, others left open (but always aiming for solutions), and many calculations had to be done. This long process of trial and error finally led to the invention of calculus. A complex set of interactions between various levels of conceptual, physical, mathematical, and even theological ideas at last produced the result, the foundation of classical mechanics, the starting point of modern science.

To illustrate this arduous path, let us focus on its last phase, when mechanical investigations, still strongly embedded in a philosophical context, made a heroic effort to overcome conceptual inertia and adapt themselves to empirical data.

2. SCIENCE AND ARTILLERY

As we remember, the prehistory of the science of motion goes back to Zeno's problem of the flying arrow. Unfortunately, however, as time passes applications of science become more and more dangerous. In the first half of the 16th century Nicolo Tartaglia studied the motion of cannonballs and substantially contributed to the more efficient use of artillery. He initially had scruples about whether he should publish his discoveries since "it would be a most blameworthy thing to teach Christians how they could better slaughter one another."[1] However, when in 1537 the Turkish invasion was imminent, his scruples evaporated, and his work *Nova scientia* was published. It certainly deserves an honorable mention in the history of mechanics. Alexander Koyré was thinking precisely about this work when he wrote that sciences are usually born from false theories.[2] Tartaglia's theory was false, but the way he posed the problem marked a new chapter in the science of motion.

Tartaglia does not try to free himself from the burden of Aristotelian doctrine. He seems to accept it without any discussion, and does not enter into subtle analyses of natural places and the nature of motion; instead, he struggles to find some quantitative and geometric characteristics of motion. His work is not simply one more treatise *de motu*, but the beginning of a new approach, *nova scientia*, indeed. Tartaglia does not avoid practical conclusions referring to the accuracy of artillery fire but, in general, he keeps his analyses on an abstract level. Any artificially made machine able violently to throw heavy bodies into the air is for him a "moving factor," and the "heavy bodies thrown into the air" are to be understood as spherical bodies made of lead, iron or stone, or of any other material similar to them as far as their heaviness is concerned.

Tartaglia discussed, among other topics, the form of the trajectory of a cannonball. Aristotelian theory, then commonly accepted, taught that natural motion cannot be mixed with enforced motion. According to this theory a ball fired by a cannon first moves along a straight line, and only later, when the impetus given to the ball by the initial explosion vanishes, does the ball fall, again along a straight line, towards the center of the Earth. Thus the trajectory of the "projectile" consists of two parts, both of them being segments of straight lines. Tartaglia saw the artificiality of this solution, and proposed a compromise: both straight segments are connected by a curved segment. The Aristotelian broken trajectory became continuous.

This is an instructive example showing how difficult is to see something that is excluded by the accepted theory. Tartaglia's solution is false, evidently contradicting simple experiments (provided one would have enough courage to perform them in a critical way), but it could be regarded as a "first approximation" to the correct solution. However, Tartaglia had enough courage to propose the "second approximation." In his *Quesiti et innovationi diversae* he admits that the trajectory of a projectile is curved along the whole of its length. The misleading theory excluding "mixed motions" has been rejected. Since, however, as the history of science teaches us, a bad theory is better that no theory, Tartaglia had to invent a new one. His new theory stated that when a body looses velocity, it gains weight.

Quesiti is written in the form of a dialogue. When Prince Francesco d'Urbino protests against this new theory ("everybody knows that at least a part of the trajectory of a projectile is a straight line"), Tartaglia answers him that the weakness of the human intellect permits us only with difficulty to distinguish the true from the false. Life seems to corroborate this remark. An artilleryman once asked Tartaglia at which angle a cannon would shoot farthest. Tartaglia answered that the correct angle was 45 degrees, but military experts objected that that was too high. However, in scientific matters the last word belongs to the experiment: "those who backed Tartaglia won their bets at an experimental test."[3]

In the 16th century Tartaglia's theory, expounded in his *Nova scientia*, enjoyed great popularity, but its revised version from the *Quesiti* was simply ignored. Long after its publication artillerymen aimed cannons by directing their barrels straight onto the target.

3. FALLING STONES

Simplifying problems is an important aspect of the scientific method. The problem of the shape of a projectile's trajectory can be altogether eliminated if one considers a freely falling body. One can then focus on another key problem of mechanics, the velocity of a falling body.

According to Aristotelian doctrine, the velocity of falling bodies depends on their weight: heavier bodies fall quicker since their tendency toward the natural place (the Earth's center) is greater. The falsity of this doctrine is especially malicious. Seemingly obvious "sense data" skillfully imitate experimental results. This conviction lasted many centuries with no opposition from any contrary opinion. The first to raise an objection was Gianbattista Benedetti, a disciple of Tartaglia and teacher of Galileo.

Already in the Dedication Letter opening his early work, *Resolutio omnium Euclidis problematum*, Benedetti declares that in this book he will destroy Aristotle's theory that heavy bodies fall with greater velocity that light ones. However, the error was so deeply rooted that it would have been impossible to destroy it with the "first blow." Benedetti was not able to free himself from intuitions connected with the concept of impetus. He claimed that only bodies "of the same nature" fall with the same velocity regardless of their weight. He probably wanted to say that it is the ratio of the weight of a falling body to the weight of the medium that determines the velocity of the fall.

It is interesting to look at a given historical process from the perspective of somebody who knows the solution. It would seem that Benedetti was facing a purely experimental question, but we have already seen that it is practically impossible to separate an empirical question from theory. Moreover, experimental results, together with the theory behind them, create what could be called a "problem situation." Its main constituent is a net of concepts, usually supplied by a theory, which determine both the formulation of the experimental question and the interpretation of the results obtained. The concepts are not static; they evolve together with problems; they determine the riddle that must be solved, and are modified by requirements of the solution.

The concept of acceleration was known at least from the time of Nicolaus Oresme who tried to make it precise in his theory of the "latitude of forms," but it was still far removed from the clarity it has today, and the ideas of instantaneous velocity and instantaneous acceleration were simply beyond conceptual reach without help from calculus. And without these two concepts, the practical applications of many mechanical problems had to remain immersed in manifold inconsistencies.

The concept of weight created even greater difficulties. It was understood in an intuitive way (although there were some attempts to make it more precise), and in thinking about it several different elements were mixed together: tiredness as felt by our muscles when carrying heavy bodies, the tendency of such bodies to the Earth's center, and the philosophical idea of the "quantity of matter." First of all, there was no clear consciousness that concepts should be defined in such a way that to each of them a "quantity" would correspond that could be numerically determined by experiment. Only when all these intricacies have been smoothed out, and the "problem situation" becomes mature enough, will experiment be able to confirm that all bodies, independently of their weight and composition, fall with the same acceleration. Tartaglia and Benedetti are but steps in the process which will soon be completed.

It is interesting to notice that Benedetti in his works wanted to imitate Archimedes' geometric method. However, even here there was a hidden trap. Archimedes' method was fruitful in statics and hydrostatics but, too literally transferred to mechanical problems, it led Benedetti to the false conclusion that the velocity of a falling body depends on the ratio of the body's weight to the weight of the medium. On another occasion, however, the Archimedean method suggested a valuable result. When Benedetti analyzed, with the help of geometric tools, the erroneous claims of Aristotle concerning rotational motion, he came to the conclusion that a rotating body, equipped with an impetus, "wants" to continue its motion along a straight line. It was a step towards the formulation of the Law of Inertia, without which mechanics is impossible.

4. GALILEO THE RELATIVIST

When we speak about the laws of motion, we must, sooner rather than later, come to the principle of inertia, and in physics this principle is inseparably connected with the name of Galileo. When, in turn, we speak of Galileo, we cannot avoid mentioning his battle with the geocentric system. The essence of the Copernican revolution can be seen in the transition from the geocentric to the heliocentric reference frame. And as such it could be regarded as a "relativistic" intervention. However, Copernicus himself was not able to go beyond the idea of absolute motion. This can be seen in his answer to the argument that Ptolemy and his followers developed against the motion of the Earth. The argument states that, if the Earth were in motion, all bodies floating in the air would constantly move to the west. "For the earth would always outstrip them in its eastward motion."[4] Copernicus' answer was the following. If the Earth is moving, its motion should be regarded as natural. What is natural cannot produce violent effects. Therefore, no violent effects can be observed on the Earth.

> But things which are caused by nature are in a right condition and are kept in their best organization. Therefore Ptolemy had no reason to fear that the Earth and all things on the Earth would be scattered in a revolution caused by the efficacy of nature ...[5]

The only possible way for Copernicus to have been freed from the objection of a purely verbal argument would have been to agree with the following principle: no observer on the surface of a body in a state of natural motion, can perform any experiment which could decide whether the body is moving or not. This principle is generalized, as compared with the text of the *De revolu-*

tionibus, in two respects: first, the principle speaks of any motion, whereas Copernicus himself was interested only in the motion of the Earth; second, Copernicus spoke of "effects due to force," whereas the principle which we have formulated on his account, speaks of any experiments performed on bodies in natural motion.

The principle which we ascribed to Copernicus was explicitly formulated by Galileo, with no reference to natural motions.

> *Then let the beginning of our reflections be the consideration that whatever motion comes to be attributed to the earth must necessarily remain imperceptible to us and as if nonexistent, so long as we look only at terrestrial objects; for us inhabitants of the earth, we consequently participate in the same motion.*[6]

This is valid not only with respect to the motion of the Earth. What Galileo wants to tell us is that no experiment performed within a mechanical system can inform the experimenter about its motion. For merchants making their voyage on a vessel, the movement from Venice through Corfu, Crete, and Cyprus to Aleppo is "as if nonexistent."[7]

Galileo continues to employ the old term "impetus;" however, he changes its meaning. The idea of impetus was introduced by Buridan, the Paris master of the 14th century, as another attempt to explain the phenomenon of the flying arrow in spite of the fact that no visible force is acting on it. The force of the bowstring imparts to the arrow a fluid-like impetus that drives the arrow forward until it is totally exhausted. For Buridan, impetus is a kind of efficient cause and as such distinct from the moving body. For Galileo, on the other hand, impetus is identical with motion itself. In order to be continued, motion has no need of any "extrinsic agent." Motion is not an accident or property of a moving body but a state of a body. Rest is just a particular case of such a state.

Koyré characterizes Galileo's physics as a physics of heavy bodies, in contrast to Descartes' physics, which was that of colliding bodies,[8] and that was why Galileo could not identify inertial motions as uniform ones. There are no "eternal" uniform motions (along an infinite straight line), because bodies are forced to move in circles by other gravitating bodies. We can find in Galileo an ingenious argument supporting this view:

*I conceive a body, launched on a horizontal plane, and by effort of thought I
assume all impediments to be removed. It is clear from what has already been
said that its movement on the plane will be uniform and perpetual if the
plane extends infinitely. But if we conceive the plane as limited, the body
(which I take to be endowed with gravity) will arrive at the end of the plane
and will continue forward, having in addition to the former uniform and
nondescending motion, that of the descent proper to its gravity. So that the
resultant is a motion composed of a horizontal uniform motion and a motion
of vertical descent, uniformly accelerated.*[9]

The addition of these two motions may give a closed orbit. In this way, according to Galileo, the law of inertia is responsible for the circular motion of planets. We can see here something of the Aristotelian unwillingness to make idealizations. If Galileo's "effort of thought" had been a little more audacious, if he
had dared to assume that the movement on the plane could continue indefinitely, he would have discovered the First Law of Dynamics. In Tannary's
opinion, the principles of Galileo's mechanics "were, it appears, an engine of
war designed to defend the Copernican view."[10] Galileo at this period was
more interested in winning the war than in creating a new physics.

5. THE GREATEST DISCOVERY OF ALL

Galileo made real progress when he stopped fighting Ptolemeans and Aristo-
telians and applied his principle of inertia to concrete scientific problems. He
did this in his later work, *Discorsi intorno a due nove scienze*,[11] in which he
developed the correct theory of a material point moving under the action of a
constant force. Such a motion contains two cases: that of uniform motion and
that of uniformly accelerated motion. The former is the simplest of all possible
motions, and the latter is the second in this respect and is a "natural motion"
of free-fall. Although Galileo never used the name "principle of inertia," the
fact that he applied this principle to his theory of uniform motion makes him
its discoverer. The theory of free-fall is also of the utmost importance because
a freely falling body is (at least from our present point of view) a "clinical case"
of motion under the action of a force, which can be studied in isolation from
the "rest of the universe." We are justified in saying that these two cases are the
cornerstone of classical mechanics.

Galileo's principle of inertia has an important consequence. If we are unable to distinguish uniform motion from the state of rest, this means that no force is acting. In other words, *if no force is acting, the body remains at rest or moves uniformly.* And this is what later on will be called the Second Law of Dynamics (or the Principle of Inertia). Galileo looked for its empirical verification, and found one in a very simple thought experiment. A little ball is rolling down along an inclined plane; of course it moves with acceleration. To push the ball uphill we must use a force, and if this force is not too great, the ball will move with deceleration. Moreover, if the angle between the plane and the horizontal direction is arbitrarily small, the force which has to be used can also be made arbitrarily small. The conclusion is that if the plane is in the horizontal position, no force should be used and the ball will move uniformly (without any acceleration or deceleration).

Koyré[12] remarks that Galileo introduced, without being aware of it, a new ontology of motion. In the Aristotelian approach motion was a *process*. A process develops if it is driven by a "moving factor;" if the "moving factor" ceases to act, the process stops. In the Galilean approach motion is a *state*. No cause is needed to support a state. A body continues to be in a given state as long as something external does not force it to change that state.

The great efforts of a generation of thinkers were necessary to understand the essential aspects of the motion of a freely falling stone. This was an extraordinarily fortunate choice. The majority of natural phenomena would never have surrendered to even the most ingenious minds. Strictly speaking, in the universe there are no isolated phenomena. Everything interacts with everything. Phenomena and processes form a hierarchy of networks of various feedbacks and dependencies. The whole is too subtly interwoven to easily disclose its secrets. One can only try desperate linguistic measures to express in words the richness of everyday experience. It is in this way that over the centuries people attempted to understand the world, but in vain. Linguistic measures only simulated understanding. A spark of genius was necessary to spot a phenomenon that could be isolated from the rest without damaging it too much, and that could be rich enough with information to convey a broader knowledge, not restricted only to itself. A freely falling stone is such a phenomenon. We can truly say that it is the understanding of this phenomenon that created modern physics. In this sense, it is the greatest discovery of all.

However, we should be aware that even Galileo's genius would have been useless if the world had not possessed the surprising property that its structure can be approximated by simpler structures. It is not an *a priori* necessity. We could easily imagine a world deprived of this property. In such a world, the researcher would have two options: either to surrender from the very beginning, or to face the world in all its complexity. If the researcher resisted the first option, the only possibility would be the strategy of a purely verbal taming of nature.

Chapter 12

~

BIRTH OF THE METHOD

1. THE VIEW FROM THE SHOULDERS OF GIANTS

*I*t is commonly accepted, not without reason, that the birth of modern physics coincides with the date of Isaac Newton's *Philosophiae Naturalis Principia Mathematica*. It was Newton himself who once said that he was able to see further than others because he was standing on the shoulders of giants. It is true that the giants, like Copernicus, Kepler and Galileo, had prepared the way and had constructed a great part of the edifice, but it was Newton who sharpened the method, formulated the laws and laid the foundation for future developments.

The best way to begin a chapter on Newton's achievement is to just open his *Principia* and immerse ourselves in reading. Newton's masterpiece consists of three books. The goal of Book One is to formulate the laws of motion. In Book Two the laws are applied to various particular cases; it is here that the foundations are laid down for new chapters of physics: the mechanics of fluids, the motion of bodies in different media, wave mechanics, and so on. And finally, in Book Three, Newton constructs his "system of the world." He deduces Kepler's laws of planetary motions from the principles of his dynamics, and develops his theory of universal gravity.

In this monumental work the contemporary reader, armed with all modern methodological tools, can easily distinguish three levels: (1) the mathematical level, which consists in a mathematical analysis of the laws of nature; (2) the physical level, in which the empirical consequences of the laws of nature are explored; and (3) the philosophical level, in which Newton struggles to find "causes" of the natural laws. In our contemporary textbooks of physics, levels (1) and (2) are present, usually in the form of a mathematical formalism and its physical interpretation. Level (3) is today often eliminated or reduced to its minimum. In Newton's work, on the other hand, it plays the essential role both in the composition of the work and as far as his own views were concerned.

2. DEFINITIONS AND LAWS OF MOTION

Having completed this quick tour through Newton's *magnum opus*, let us start from the very beginning. After the Dedication Letter and the Author's Preface, Newton starts with a section entitled *Definitions*. Without any introductory explanation he offers the reader the first definition:

> *The quantity of matter is the measure of the same, arising from its density and bulk conjointly.*[1]

This is not a modern text, so we must learn how to read it correctly. Newton wants to tell us that the "quantity of matter" is something that can be measured, and that the "measure" is obtained by multiplying density and volume ("bulk"). He does not try to explain to the reader what is the essence of matter, he is only interested in its "quantity," and he gives a prescription of how to measure it. Such definitions are called today *operational definitions*.

A few lines below Newton explains that this definition is based on an idealization: "I have no regard in this place to a medium, if any such there is …," and he proposes to replace the old term "quantity of matter" by a new one: "It is this quantity that I mean hereafter under the name of body or mass." The latter term is now commonly used in physics. When reading the above text of Newton we are witnessing the birth of its purely operational meaning. And if we agree with the majority of philosophers of science that only operationally defined concepts deserve to be called physical concepts, we are in fact witnessing the birth of modern physics. There are only very few books in the world whose first sentences would be of such great weight.

Let us now quote the third definition:[2]

> *The vis insita, or innate force of matter, is a power of resisting, by which every body, as much as it lies, continues in its present state, whether it be of rest, or of moving uniformly forwards in a right line.*

Newton is still using a traditional term "innate force of matter," but endows it with the new meaning. He explains:

> *This force is always proportional to the body whose force it is and differs nothing from the inactivity of the mass, but in our manner of conceiving it. A body, from the inert nature of matter, is not without difficulty put out of its state of rest or motion.*

This justifies the new name: "Upon which account, this *vis insita* may, by a most significant name, be called inertia *(vis inertiae)* or force of inactivity." The term "inertia" is now commonly accepted. It is a force that a body exerts only "when another force, impressed upon it endeavours to change its condition." A clear definition pays off by elucidating other concepts. The principle of relativity turns out to be just a corollary of the above definition of inertia:

> *Resistance is usually ascribed to bodies at rest, and impulse to those in motion; but motion and rest, as commonly conceived, are only relatively distinguished; nor are those bodies always truly at rest, which commonly are taken to be so.*

The next step in this logical chain is, of course, the definition of force. Newton's fourth definition reads:

> *An impressed force is an action exerted upon a body, in order to change its state, either of rest, or uniform motion in a right line.*

And the explanation:

> *This force consists in the action only, and remains no longer in the body when the action is over. [...] But impressed forces are of different origins as from percussion, from pressure, from centripetal force.*

The third definition has established the standard of motion, and the fourth definition identifies force by a deviation from this standard; it becomes a truly operational definition only together with the Second Law of Dynamics. But with these definitions, the laws of dynamics ("Axioms, or Laws of Motion," as Newton calls them) are almost obvious:

LAW I: Every body continues in its state of rest, or of uniform motion in a straight line, unless it is compelled to change that state by forces impressed upon it.

LAW II: The change of motion is proportional to the motive force impressed; and is made in the direction of the straight line in which that force is impressed.[3]

From our previous analyses we know that, although formally speaking, the First Law is the consequence of the Second, it is not superfluous, because it establishes the standard of motion without which the Second Law would be meaningless.

3. CALCULUS

In Newton's formulation of the Second Law there appears a crucial expression: "the change of motion." Zeno's paradoxes have demonstrated how many intricate difficulties are involved in the concept of motion, and here we have not only motion but also its change. In our modern formulation we would say "acceleration." Aristotle and many of his followers were able correctly to use the concept of average velocity. Galileo quite efficiently dealt with uniformly accelerated motion. But in order to speak meaningfully about instantaneous velocity and instantaneous acceleration in general, calculus is needed. It had been discovered independently by Leibniz and Newton, and became a source of a heated rivalry between them.

As with all great scientific discoveries, especially in mathematics, the discovery of calculus was anticipated by a long chain of partial results and struggles with various problems. In the 17th century, ideas suggesting the correct solutions were "hanging in the air." For instance, the philosopher Thomas Hobbes, who was considerably impressed by Galileo's mechanical achievements, postulated that motion should be made the basis for the whole of natural philosophy. To implement this idea he introduced the concept of *conatus* (tendency, inclination), that would play the same role in the analysis of motion as the concept of point does in the science of extension, i.e. in geometry. Just as an extension consists of points, motion is supposed to consists of *conatus*. He regarded time as a pure *phantasm*, and reduced it to the "before-and-after" in motion. *Conatus*, he argued, should be understood as motion in an infinitely small portion, i.e. as small as we are able to conceive.[4]

Newton's teacher, Isaak Barrow, sympathized with Hobbes' views but, in contrast, thought that the concept of time was crucial in the analysis of motion. Although "time does not imply motion, as far as its absolute and intrinsic nature is concerned[5], [...] time implies motion to be measurable; without motion we do not perceive the passage of time." For Barrow time was more a mathematical than a physical concept, having many similarities with geometric line, "for time has length alone" and "can be looked upon as constituted from a simple addition of successive instants or as from a continuous flow of one instant ..." Whitrow justly notices that for the first time we meet here a clear idea of the geometrization of time.[6]

In his *Lectiones geometricae* Barrow went as far as to propose a method of finding the tangent to a curve with the help of calculations which are very close to our geometric interpretation of derivative. He was thus not far from the great breakthrough, but before sending his *Lectiones* to be printed, Barrow gave the manuscript to Newton for checking and making final corrections, and he then abandoned mathematics to immerse himself in theological studies.

Both Newton and Leibniz (as we know today with no shadow of a doubt, independently of each other) defined the fundamental concepts of calculus in such a way that they became effective tools of calculation. Both of them produced a host of formulae on how to differentiate and integrate various functions, and both of them understood that differentiation and integration are operations inverse with respect to each other, but neither of them had a clear idea of the limit function, and this was the main reason why calculus had to wait several generations until the work of Cauchy, Cantor, Dedekind and Weierstrass to obtain its firm foundations. Newton based his definitions on the intuition of a "flow" of motion (he even called his approach the "calculus of fluxions"), whereas Leibniz was inclined to regard motion as consisting of arbitrarily small, indivisible parts (we find the trace of this in our present notation, df, as the differential of a function, which we owe to Leibniz). However, the results obtained by both Newton and Leibniz were so similar that to resolve their quarrel about priority was not easy.

There is no doubt that when preparing his *Principia* Newton made much use of calculus, but he presented the results in a purely geometric way. Only in a very few places can the diligent reader of the *Principia* find some traces of differentiation. In this respect Newton followed Barrow's style, and most probably wanted, by sticking to the traditional method, to increase the potential number of his readers.

Greek antiquity initiated two streams of investigations. Zeno's paradoxes gave birth to mathematical analyses of motion, and with Aristotle's physics dynamical research began. And only when these two lines of investigations met, could modern physics take off. This happened in Newton's work.

4. CONCEPTS WELL KNOWN TO ALL

From the story told in the preceding chapters we have drawn the lesson that the evolution of concepts is one of the most important driving forces of scientific progress. We also know that concepts do not evolve alone, but participate in a more complex process of solving problems. The problem of motion has matured to be solved in Newton's work, and the concepts related to motion have matured together with it. The crucial point is that from now on concepts in physics must be defined operationally, that is to say, in such a way that, through measurement procedures, they can be changed into numbers. Newton was fully aware of this when he collected his definitions into the first section of the *Principia*. Concepts defined here enter into the very core of his mechanics.

But a long tradition accumulated some other concepts around the science of motion which, as it has turned out, do not directly have an influence on the laws of dynamics. Newton does not avoid discussing them, but separates them from the main body of his physics. After the section on Definitions there follows a section entitled *Scholium* devoted to what we would today call the philosophical aspects of Newtonian dynamics. Its opening paragraph reads:

> *Hitherto I have laid down the definitions of such words as are less known, and explained the sense in which I would have them to be understood in the following discourse. I do not define time, space, place, and motion, as being well known to all. Only I must observe, that the common people conceive*

94

those quantities under no other notions but from the relation they bear to sensible objects. And thence arise certain prejudices, for the removing of which it will be convenient to distinguish them into absolute and relative, true and apparent, mathematical and common.

In fact, the definitions laid down in the preceding section of the *Principia* were not "less known" but completely unknown, at least as far as their operational character was concerned. The ideas of "time, space, place, and motion," although "well known to all," should be explained in order to eliminate "certain prejudices." And this is precisely the aim of the *Scholium*. We find here the famous "definitions" of "absolute, true and mathematical time," absolute space and motion, and their relative counterparts as well. They are not defined operationally, and that is why they belong to the third, philosophical level of Newton's work. That is not to say that they are not important. On the contrary, they were vital for Newton himself, and they have initiated a new stream of thinking for many generations of physicists and philosophers.

5. THE ELIMINATION OF MATTER

The concept of matter has a long history[7] but, strangely enough, was only relatively recently understood in the sense close to our current understanding of the term. In Aristotelian physics the term matter, usually specified as *primary matter*, denoted something almost totally "immaterial," namely the pure potentiality of receiving forms. The materialistic monism of the ancient atomists doubtlessly helped to consolidate the abstract ideal of materiality in our philosophical consciousness. In antiquity two tendencies dominated. One, following Euclid's "geometric" definition that "a body is what has length, width and depth," identified the essence of being a body with extension. The second, continuing the Stoic tradition, connected the idea of a "material body" with an inertia understood as a lack of any activity (an echo of Aristotle's prime matter concept). The first of these tendencies finally led to the Cartesian doctrine of the identity of matter with extension; the second tendency to the medieval disputes concerning "quantity of matter" and the impulse *(impetus)* of Buridan.

The term "quantity of matter" *(quantitas materiae)* belongs to Aegidius Romanus, a disciple of Thomas Aquinas, who in his reflections on the Eucharist taught that there must exist a certain fundamental property of matter, a kind of substratum that, after transubstantiation, sustains accidents of bread and wine when their substances cease to exist. This property, the closest to substance, he called *quantitas materiae.* The argument he put forward was ingenious (and entirely non-theological): when one changes a volume, the "quantity of matter" remains the same, in spite of the fact that density changes as well (the first ever "conservation law"!).

It is not an accident that such speculative ideas opened the door to mathematical inquiries. Let us notice that the most fundamental property, the one "next to substance," turned out to be a *quantity* (of matter). It was Buridan who formulated the law that the impetus is proportional to the quantity of matter which, although erroneous, transferred the quantitative point of view from matter to motion. Kepler continued this line of thinking, enriching it with observational aspects. He thought that planets have different "quantities of matter;" therefore, by observing along which trajectories and with which velocities they move, we can verify our hypotheses concerning their laws of motion.

The next step belonged to Newton. As we have seen, he was still using the term "quantity of matter," but in a new, precisely defined meaning. In fact, in the first and second levels of his work (mathematical formalism and physical interpretation) he initiated the process of eliminating the matter concept from physics. Indeed, the concept of matter is not a physical concept since it is not defined in an operational way. One can measure mass (Newton has taught us how to do this), energy, volume and density, but one cannot measure matter. The concept of matter, often used by physicists, is taken either from a philosophical vocabulary or, more often, from everyday language (we should not forget that the latter is usually contaminated by philosophical meanings). In Newton's work, mass is a fully operational concept, something that can be measured, something about which we know only through measurement. The result of measurement is a number and a number, when represented in an algebraic formula, becomes just a symbol. In this sense, in the formal structure of Newtonian mechanics, mass has been reduced to the role of a parameter, usually denoted by m.

Newton himself probably did not notice that, in the mathematical and physical levels of his work, the concept of matter had been eliminated and effectively replaced by that of mass. But it was left intact at the philosophical level. Moreover, there it played a key role. Pre-Newtonian philosophical tradition distinguished *primary properties,* to be ascribed to every sensual object; and *secondary properties,* to be ascribed only to a certain class of sensual objects. Newton, in his philosophical speculations, understood matter as a substrate of primary properties, thus as something very close to the traditional idea of substance. In Newton's time four properties were regarded as primary:

extension, impenetrability, movability (by something else), and inertia (understood as "passivity"). Newton added to this list two more properties: the ability to attract gravitationally and the ability to be attracted gravitationally. In the Aristotelian tradition matter was totally passive, requiring "something external" to change and to move. By ascribing to matter the active property of being able to attract by gravity, Newton unconsciously initiated a long historical process which finally led to the birth of modern materialism, the doctrine that matter is an "active principle" of everything. [8]

6. TO CALCULATE OR TO EXPLAIN

In René Thom's view Newton explained nothing, but calculated everything.[9] It is true that it was Newton who, for the first time to such an extent, introduced the method of mathematical modeling of various phenomena, but it is not true that he was not eager to supply explanations. In fact, he was rather unhappy that his theory of gravity was unable to identify any cause of gravitational attraction, and considered four different hypotheses which could provide such an explanation. According to the first hypothesis, cosmic ether was responsible for the propagation of gravity; according to the second hypothesis, this role was played by light. The third hypothesis was more mysterious, making recourse to "active principles" that were supposed to fill in the space and be of a non-mechanical nature. The fourth hypothesis ascribed gravitational effects to the direct action of God. Only much later, when philosophical categories were eliminated from science, did many authors see in Newton a precursor of the positivistic style of thinking.

Starting from Newton, there begins a period in which physics is totally conquered by the method of constructing mathematical models and deducing from them predictions that could be compared with measurement results. But again, it is not true that such models explain nothing. A mathematical model of a physical phenomenon is not just a device for computing various measurable effects; it also gives an insight into the "inner structure" of the phenomenon, and this must be termed understanding.

The great success of modern science firmly justifies the conviction that there exists a certain similarity between the mathematical structure of a given model and the structure of the modeled phenomenon. There is, in a sense, a resonance between them: on the one hand, the empirical results verify the mathematical model and, on the other hand, the model allows one to interpret the empirical data, and to project new experiments. We are then entitled to say that the structure of the model reflects or represents the structure of the aspect of the world under investigation. Moreover, no mathematical structure is isolated from other mathematical structures; there exists a complex network of interactions between them. If we mathematically model a certain physical phenomenon, we automatically put it into a multiform net of interactions with other mathematical structures. And immersing a phenomenon into a broader context of logical deductions makes it more intellectually transparent, i.e. more understandable. Therefore, the empirical-mathematical method of investigating the world is by no means only a computational device; it also gives us an understanding of the inner functioning of the world, possibly the only authentic and viable understanding.

7. EXPERIMENTAL PHILOSOPHY

There are reasons to believe that Newton himself was not fully aware that he was creating a new science, distinct from the philosophy of nature or natural philosophy of the period. He knew, to be sure, that his discoveries were of the utmost importance and were opening new perspectives for the future, but he believed that he still was doing time-honored natural philosophy. The title of his main masterpiece clearly testifies to this. Both in Newton's writings and in the rich literature which soon abundantly grew around his achievement, some other terms were also used, such as "experimental philosophy" or "mechanistic philosophy." In Newton's writings these have no strictly determined meanings, but the use of the qualification "philosophy" clearly indicates that the process of splitting philosophy into particular scientific disciplines was beginning. Moreover, the reading of the whole of Newton's work, and some of his statements, allows us to formulate a rough idea of what he had in mind when using such expressions. For instance, in one of his manuscripts we read that the task of natural philosophy is to discover the scheme of nature's operating and reduce it, as far as possible, to general rules or laws which, however, should be established with the help of observation and experiment.[10] Or in a letter to Cotes, who edited the second addition of the *Principia*, Newton explains that natural philosophy, starting from phenomena, formulates, by induction, general statements.[11] The precise meaning of these, and many similar, statements is the subject of prolonged discussions among specialists, but it could hardly be denied that Newton was fully conscious of the novelty of his method. We could guess that for him "experimental philosophy" was a new method rather than a new branch of science or even a collection of new results. But the results were conspicuous and quickly accumulating, both in their number and in their quality. A new image of the world began to emerge out of them – an image so powerful that very soon it became the entire content of the term experimental or mechanistic philosophy[12].

Chapter 13

~

IS THE WORLD MATHEMATICAL?

1. THE METHOD

*A*t the end of Part One, in Chap. 5, we proposed a hypothesis that the world is rational, in the sense that it has a property owing to which it can be rationally investigated. There are many methods with the help of which the world can be investigated but, as we have seen in our laborious pathway through the adventures of human thought, only one of them has proved to be especially efficient, namely the method of constructing mathematical models of various aspects of the world and of checking them experimentally. When physics started using this method on a large scale, the progress in understanding the world became so rapid that it cannot be paralleled by the progress in any other field of human activity. This fact allows us to express our hypothesis more precisely: we should ascribe to the world a property owing to which it can be efficiently investigated with the help of the mathematical-empirical method (for the sake of brevity, in the following pages we shall call it simply the mathematical method). In this sense, we will often speak of the *mathematical rationality* of the world, or simply say that the world *is mathematical*.

Here we should make two remarks: first, the fact that we sometimes omit the term "empirical" in the name of the scientific method is not intended to minimize the role of experimentation in science. Without performing experiments we would have a game consisting of constructing various mathematical models rather than any investigation of the world. Second, we must firmly emphasize that without the strong "contamination" of all experiments with mathematics, experiments would simply be unthinkable. This refers to all kinds of experiments: from the simple ones performed by Archimedes to the most sophisticated ones performed by big modern elementary particle accelerators. Some radical rationalists claim that all information about the world could be deduced from some fundamental theory if only some day we could find it. Even if this were true (in fact, we do not think it is), experiments would be indispensable in order to check whether the theory was correct. Moreover, by focusing on the mathematical-empirical method we do not want to underestimate other methods of cognitive and emotional contact with the world. We are interested simply in the physical world, and in that domain the method has no rival.

2. NON-MATHEMATICAL UNIVERSES

Some people claim that to say that the world is mathematical is trivial. Among physicists and mathematicians it is usual to call a statement trivial if it is conceptually empty or sterile of information. To say "X is trivial" means that X tells us nothing new. In order to show that this is not the case as far as our hypothesis of the mathematical rationality of the world is concerned, we should decide whether the concept of a non-mathematical world is self-contradictory or not. If it is, our hypothesis is indeed empty (tautological). To show that this is not the case, we shall proceed in the following way.

By using the method of "gedanken experiments," one can construct various non-mathematical world models, i.e. world models devoid of the property owing to which they could have been investigated with the help of mathematical method. The examples of such non-mathematical models form a hierarchy: from a non-mathematical world in the stronger sense to non-mathematical models in the weaker sense.

Let us begin with the "maximally non-mathematical world." It would be a universe in which no laws of mathematics and logic are obligatory, not only the laws of *our* mathematics and logic but of *no* mathematics and *no* logic whatsoever. Let us make this clear: the laws of probability are also excluded in such a world (probability theory is as good a mathematical theory as any other mathematical theory). We shall call such a world the *Non-Mathematical World 1* (*NMW1*, for brevity). In this kind of world there are no regularities, or, equivalently, all possible regularities are simultaneously valid: everything is allowed to happen. We now make an *ontological hypothesis* that the *NMW1* cannot exist. Its contradictory character (everything in it is allowed) excludes it from existence. As contradictory, *NMW1* is also irrational; to be rational a universe must be, at least in some sense, mathematical.

Our present knowledge of mathematics allows us to imagine a universe the structure of which would correspond to mathematical structures that would surpass our cognitive possibilities. In the historical development of mathematics a powerful selection effect is acting: we investigate only those mathematical structures which we can investigate. We know, for example, that there exist many mathematical functions which are too complicated to manipulate or even to express in a formula. In fact, the vast majority of functions belong to this "exotic" (from our point of view) category.

As an example, let us consider a highly idealized universe which can find itself only in two states: state "zero" and state "one." The history of this universe is a sequence of zeros and ones. Let us assume that our universe had a beginning which we will denote by putting a full stop at the beginning of the sequence of zeros and ones. We obtain, for instance, the following sequence

.011000101011

The goal of the physicist living in this universe[1] is to construct a theory, based on which it would be possible to predict the future states of the universe. Such a theory would consist of a formula, essentially shorter than the sequence of zeros and ones itself, which would permit us to calculate subsequent elements of the sequence. Our physicist has a chance to find such a formula only if the sequence is *algorithmically compressible*, i.e. if an algorithm can be found permitting us to compress the sequence to a shorter form. But here we have a problem!

The sequence of zeros and ones (with a full stop at the beginning) can be interpreted as a decimal expression of a real number from the interval $[0,1]$. However, it is well known that the set of real numbers from $[0,1]$ that are algorithmically compressible is of measure zero. This means that the chances of selecting such a number from the interval $[0,1]$, when choosing at random, are nil. Therefore, if the universe in question was not created by a highly mathematically minded Creator, our physicist has no chance of finding the theory of this universe. Such a universe is mathematical, but is not knowable.

Of course, the physicist could say that the sequence of zeros and ones is itself a theory of the universe. But, in the case considered, such a sequence is but a copy of cosmic history. The conclusion is that the physicist could have neither the exact copy of cosmic history nor any theory at all. The universe cannot be approximated by any simpler mathematical structures; it has no algorithmic compressibility property. *For us* such a world is not mathematical. We will call such a world *NMW2*.

Approximations and idealizations play crucial roles in science. If we were sentenced to face the world in its entire complexity, our physics would probably remain on the level of purely qualitative descriptions. The moment when Newton understood that instead of "real bodies" it is worth considering material points moving uniformly under the action of no forces was the starting point of modern physics.

There is yet one possibility. Let us imagine a universe exactly the same as ours with only one "small" exception. Let the force of gravity in this universe between two masses, instead of acting in inverse proportion to the squared distance between them, act in inverse proportion to that distance to the power 1.999. In such a universe, planets would revolve around their suns along complicated curves, in general non-closed and non-periodic ones, and if even life could evolve on one of them, the astronomers inhabiting such a planet could hardly go beyond the Ptolemaic type of astronomy, trying all sorts of deferents and epicycles. One could doubt whether in such conditions the law of gravity would ever be discovered. This is an example of a universe that is perfectly mathematical, but almost impossible to be investigated by its inhabitants. Let us call such a world *NMW3*.

3. WHAT CAN WE LEARN FROM THE EXAMPLES?

We have called the world mathematical since it has a property owing to which it can be efficiently investigated with the help of the mathematical-empirical method. This property has a meaning only with respect to the cognitive possibilities of intelligent investigators. The above examples show that there could exist universes (at least mentally, as non-contradictory objects) that have some inner mathematical structure, but that could not be investigated by intelligent beings *(NMW2)*. In our conventional terminology such universes were classified as non-mathematical, but now we will sharpen the terminology and call such universes *ontologically mathematical* but *epistemologically non-mathematical*. The universes of the type *NMW1* are both *ontologically* and *epistemologically non-mathematical*. We believe that such universes are only an abstract possibility which, in fact, is excluded from existence.

In this and in the preceding chapters we have argued that we have to ascribe to our universe two properties: its rationality and its mathematical character. This could suggest that the latter property is secondary with respect to the former one, or that the latter property is a special kind of the former. However, such a conclusion would be premature. If to be ontologically mathematical is for our universe a necessary condition of its existence (as argued above), there cannot exist a universe that could be ontologically non-mathematical. Therefore, there cannot exist a rational universe without any form of mathematicity. A universe that can be rationally investigated must be at least ontologically mathematical.

The question arises: would it be possible to have a universe that is rational but epistemologically non-mathematical? Such a universe could be rationally investigated with the help of methods different than mathematical ones. That does not seem an unreasonable option. After all, there exist sciences which do not employ mathematical methods.

4. THE NATURAL SELECTION OF PHYSICAL THEORIES

It is interesting that discussions concerning the "mathematical character of the world" polarize views to a much higher degree that other similar polemics: whereas some believe that the problem is one of the most important in philosophy, others claim that it is trivial and not even deserving of discussion. Why is this so? I think that the reason lies in the fact that the property of the world being discussed is overwhelmingly universal. If some form of mathematicity is a necessary condition of existence, there is nothing that would not have some mathematical aspect. Without a "reference point" (i.e. something that does not have it), it is difficult to notice this property. It is rather like flying in a supersonic jet in the dark – the passengers do not experience motion (everything around them moves together with them).

This diagnosis is corroborated by the fact that when opponents of our thesis argue against it, they always tacitly assume what they want to destroy. For instance, Bas van Fraassen in his book, *The Image of Science,*[2] argues in the following way. Physicists construct many theories aimed at explaining the world, and check them with the help of experiments. It turns out that during this process the majority of proposed theories are empirically inadequate.[3] There remain fewer and fewer theories that have the chance of being empirically adequate. Finally, only one theory becomes an "obligatory theory." Something similar to natural selection occurs here. In this way, the effectiveness of mathematics in modeling natural phenomena is deprived of its supposed mysteriousness. Since scientists selected the theory that would be the most successful, it is not surprising that the theory is the "best one" in this respect. But anyway, most probably the present "best theory" will sooner or later be replaced by an even "better theory," the one which will be more empirically adequate than the present one.

Van Fraassen is essentially right that a kind of natural selection acts in the process of the competition of physical theories. But why is this possible? In non-mathematical universes, discussed in the previous sections, it would be impossible. Moreover, all selection effects are of a probabilistic character, and if we notice that the calculus of probabilities is as good a mathematical theory as any other mathematical theory, we are again facing the same problem: why is the universe mathematical?

As we have seen, this question is an aspect of the broader question, why is the universe rational? In the case of this question it is even easier to see that any argument aimed to show its triviality becomes a vicious circle; it assumes what it wants to prove. In the ontologically irrational universe no arguments can function. Anyone who wants to prove something or to argue on behalf of something tacitly assumes that the universe is not ontologically irrational.

5. THE JUSTIFICATION OF INDUCTION

It is a remarkable fact that as long as deductive methods that are logically reliable[4] were used in the sciences, progress was slow and laborious, but as soon as the experimental method, which is logically unreliable, was used, progress became rapid and spectacular. This fact does not compromise deductive methods as such (which are irreplaceable in the formal sciences, e.g. in mathematics); it simply negates using deduction as the only method of doing science. Why then have logically unreliable methods turned out to be so fruitful in the sciences? What guarantees their extraordinary effectiveness? Which additional assumptions have to be made to change this method into a logically reliable one?

This problem is known under the name of the justification of induction. Roughly speaking, the inductive method, or simply induction, is the process of inferring a general statement, or a "law", from particular instances that have been observed, measured or investigated in some way[5]. This is what, apparently, is done in the sciences: we experimentally investigate a finite number of cases (in principle as many as possible, but often in practice rather few), and generalize the conclusion to all possible cases. This strategy was often simply identified with the empirical method itself, and the empirical sciences were notoriously called the inductive sciences.

Newton himself was firmly convinced that induction was a basic method of physics, or "experimental philosophy," as he called it. He formulated his "fourth rule of reasoning in philosophy" in the following way:

*In experimental philosophy we are to look upon propositions inferred by
general induction from phenomena as accurately or very nearly true, not-
withstanding any contrary hypotheses that may be imagined, till such time
as other phenomena occur, by which they either be made more accurate, or
liable to exceptions[6].*

Let us notice that Newton is aware that the results of inductive reasoning are
only "very nearly true", and that they should be adhered to until "other phenomena occur" that will compel us to modify or to change them.

The problem of induction was for the first time clearly stated by David
Hume. He asked the question which, in Popper's formulation[7], reads:

*Are we justified in reasoning from [repeated] instances of which we have
experience to other instances [conclusions] of which we have no experience?*

Hume's answer to this question is negative: we are never justified in reasoning
from past experiences to future results, however great might be the number of
our past experiences. The argument is simple. The justification of such reasoning can be either by deduction or by induction. It cannot be by deduction since
the deductive reasoning is only justified if, by accepting the premises and de-
nying the conclusion, a contradiction is produced, and in the case of inductive
reasoning this never happens. The justification of induction cannot be done by
induction because that would be a vicious circle.

Hume's criticism of induction has ignited a long series of discussions continuing to the present day. To pursue in detail these arguments would go beyond the intent of this chapter; so we will touch upon only two issues that have
emerged during these polemics.

The first issue regards the relationship between inductive reasoning and
probabilistic inference. It appears that, based on past experiences, we are entitled to draw some *probable* conclusions about the future. One of the very
popular approaches to this aspect of the induction problem is the so-called
Bayesian inference[8]. It is based on the Bayesian interpretation of probability
which "counts" a degree of belief in a given hypothesis rather than the fre-
quency of favorable occurrences, as is the case in the interpretation of standard probability. In the procedure of Bayesian inference it is assumed that as
the evidence supporting a given hypothesis accumulates, the degree of belief

in this hypothesis increases as well. To quantify this increase, Bayes' theorem (well known in probability theory) is used.

Independent of any philosophical interpretations, probability theory is often employed in all sorts of statistical predictions in which from a relatively small subset (sample) of cases we try to draw conclusions referring to a bigger set of cases that have never been (or never will be) observed or measured. Does statistical inference solve the problem of induction? Opponents of the "statistical solution" of the induction problem emphasize that in order to apply probability calculus to make inductive inferences, we must know the initial probability distribution, i.e. the rule ascribing various "probability measures" to various events. For example, in tossing a coin there are two events: a "head" and a "tail", and to each of them we ascribe the probability measure equal to one half. Let us notice that we do that based on our long experience of tossing coins. Once we have a probability distribution we can make all sorts of calculations with the help of probability theory. And the distribution function is either assumed *a priori* (e.g. based on a "symmetry" of a given problem), or is taken from experiment, i.e. by induction[9].

The second issue is the "solution" of the induction problem proposed by Karl Popper. He claimed that the sciences do not use induction at all. In their research practice scientists formulate hypotheses (based on their knowledge, intuition, inventiveness or other contingent factors) and deduce from them consequences that could be compared to the results of experiments. Thus the standard way of constructing scientific theories is not by induction but rather by, what Popper calls, the hypothetico-deductive method. If no observable conclusions follow from a proposed theory, this theory is not falsifiable, and as such it has no place in science. If an experiment contradicts the conclusion, a given theory has been falsified, and could be of interest only for historians of science. No theory can be ultimately confirmed or verified; theories can only be corroborated if they successfully pass more and more experimental tests[10].

It is true that the inductive method finds only a limited application in the empirical sciences, but one could hardly doubt that some elements of induction form an important part of the method of science. No experiment is repeated indefinitely in scientific laboratories. After the result has been faithfully verified by a few scientific centers, it is acknowledged by the scientific community to be "universally" valid. Although scientific theories are, in general, not arrived at by induction, the problem of how to justify induction remains pressing.

Hume, in his original thinking about induction, considered the following problem. Let us suppose that we want to formulate a general conclusion from having investigated a certain number of particular cases. It can be lawfully done (i.e. in agreement with the principles of logic) if we supplement the reasoning with an additional premise. Hume noticed that such an additional premise is usually provided by the *principle of the uniformity of nature*. This principle is commonly believed in by both ordinary people and scientists. If one believes that nature "does not make jumps", and consequently that the future will always resemble the past, then one is fully justified in claiming that a "sufficient number" of cases studied can be a good representative of a more general class of cases. However, Hume was quick to add that this common practice does not solve the problem of induction, because we arrive at the principle of the uniformity of nature by induction.

After having read the present chapter the reader can rightly guess that, in a more modern approach, the principle of the uniformity of nature can be replaced by the thesis about the mathematical character of the world. Indeed, provided that, in some respects, the structure of the world is similar to a certain mathematical structure, it seems natural to claim that, if we grasp some aspects typical of this structure, we could reconstruct the entire structure. The mathematical structure itself provides the missing steps which are indispensable for logically justifying the conclusion. It is along this road that scientific progress develops. In other words, if we add the thesis on the mathematical character of the world to the set of propositions expressing the inductive method, we obtain a description of a highly reliable method.

However, we should not forget that the thesis on the mathematical character of the world does not possess any logical necessity in itself. It is only a thesis very well justified by the entire history of physics, which is nothing other than a very long-lasting process of inductive reasoning. So the problem remains. However, there is one important difference between the principle of the uniformity of nature and the thesis on the mathematical character of the world. The former is more or less an intuitive idea, whereas the latter is an assumption belonging to the very foundations of the scientific method.

Chapter 14

∼

MATHEMATICS AT WORK

1. INTRODUCTION

*T*he birth of mathematical sciences at the turn of the 16th and 17th century changed the "face of the Earth" and, in fact, opened the period called by historians *modern times*. The sciences continue to shape our mentality and our culture to an extent we even now do not fully appreciate.

In the preceding parts of our book, we traced the origin, consolidation and finally the readiness for action of the mathematical-empirical method. We argued that the great successes of the sciences are rooted in the tacit assumption that the world possesses a property owing to which it can be rationally investigated. In this sense, we say that *the world is rational*. However, it does not respond equally well to any method of investigation. As long as people tried to understand the world with the help of purely conceptual analysis, by following intuitive ideas or even by applying rigorous tools of logical deduction, starting from more or less "obvious" premises, the progress was slow, if any. But as soon as the method of constructing mathematical models and checking them experimentally had been elaborated, progress was immediate and it continues to accelerate. The rationality we should ascribe to the world is a rationality of a special kind. It is a mathematical rationality. We express this by saying that *the world is mathematical*.

In this last part of our book we will take a look at how the mathematical method works in modern science. Newton believed that he himself did almost everything that could be done with the help of the mathematical method and it was only the philosophical analysis of the results obtained that was left to his successors. In fact in Newton's work only such mechanical categories as space, time and motion (and the same for optical phenomena) were mathematized, but soon various qualities began to surrender to mathematics: colors, sounds, heat. All these qualities were reduced to mechanical quantities; in the case of heat with the help of statistical methods.

It seemed that the whole universe was but a big, pointless mechanical construct. This is how Alfred North Whitehead describes a poet's revolt against the mathematization of the world. Referring to Tennyson's poem, "The stars, she whispers, blindly run," he writes:

> *Each molecule blindly runs. The human body is a collection of molecules. Therefore, the human body blindly runs, and therefore there can be no individual responsibility for the actions of the body ...* [1]

But soon the revolt began in the heart of physics itself. The discovery of electromagnetic theory in the second half of the 19th century triggered revolutionary changes at the very foundations of physics. In the first decades of the 20th century a new physics begun to emerge, and it was by no means "mechanistically dull." If poets could follow its mathematical language, they would be delighted with its abstract fantasy, enormously surpassing any human imagination: light bending around celestial bodies, clocks accelerating and slowing down, space-times changing their geometries depending on the motion of masses, and the entirely new quantum worlds in which the material substances of elementary particles dissolve into probability waves. Even chaotic phenomena, which hitherto seemed to be reserved only for untamed contemplation, have now surrendered to the overwhelming mathematics. Physics has become poetry, but poetry that is prosaically verified in sophisticated experiments. The miracles of mathematics were not exhausted in creating general relativity, quantum physics and chaos theory. Its power is still at work. It drives us to the final goal, the full unification of physics and the ultimate understanding of the world. This is not an easy task. Sometimes it seems that this goal is at hand, just around the corner, but when one more step is taken the goal escapes beyond the horizon.

And the inevitable question: does the mathematical method of investigating the world have any limits? We can only suspect that if such limits do exist, it is the mathematical method itself that would be powerful enough to discover them. And this indeed seems to be the case.

2. MATHEMATICS SEES MORE THAN OUR EYES DO

Newton believed that he had presented his mechanics in a final form. His successors would have to provide only a few minor details. The work that remained to be done would consist of disclosing the causes of physical phenomena, such as universal gravity, but this belonged to philosophy rather than to the "mathematical description." In this respect Newton overestimated his own work. The *Principia* marked the Great Beginning, not the end.

Quite soon Newtonian physics was subject to manifold analyses. What would happen if we changed the assumptions made openly or implicitly by Newton? How to make Newton's reasoning simpler? Is it possible to improve mathematical algorithms in such a way that they would lead to quicker and more precise results? It turned out that Newtonian mechanics could be presented in the form of two different mathematical settings. Owing to this fact we have today classical mechanics in the Lagrangian form and in the Hamiltonian form. They are two incarnations of the same mechanics, but each of them suits different goals. Let us focus on the Lagrangian formulation.

Suppose we want to compute the behavior of a mechanical system, that is to say we want to know how the system will move under the action of given forces. There is a rather simple rule how to compute the so-called Lagrangian function for any mechanical system.[2] Once the Lagrangian is known, one can compute an integral which physicists call "action." And now the decisive step: the mechanical system investigated will perform only such motions for which the action assumes the extreme values (i.e. minimal or sometimes maximal values). This is a beautiful example of an exploration of the mathematical structure of the world. It is not true that classical mechanics is but a prolongation of our sensory cognition. Our senses do not tell us that the mechanical systems we observe in the macroscopic world are realizations of the "extremum principle," and it is exactly this principle that is responsible for the world of our everyday experience.

The validity of the extremum principle goes beyond classical physics. Today we know that all major physical theories can be obtained by employing the scheme invented by Lagrange. The only major problem is how to guess correctly the corresponding Lagrangian. All the rest requires only mathematical skill (which sometimes can be a great challenge): the fundamental equations of a given theory are obtained by computing the extreme value of the corresponding action integral.

The extreme action principle discloses a fundamental structural property of the world. This principle does not underline the equation of this or that physical theory but the equations of all major physical theories. All these theories differ only in the fact that each of them requires different Lagrangian functions. If we would like to create a "theory of physical theories," the extreme action principle would play in it the role of the principal axiom.

This is one more proof of the fact that mathematics sees more than our eyes can see, even if they are assisted by the most advanced inventions of electronic optics.

The process of the mathematization of motion was long and laborious but its results surpassed all expectations. Starting from Newton the progress underwent an "avalanche acceleration." Avalanches not only run on, but they also take everything along with them. The mathematization process soon went beyond the science of motion and expanded to other branches of physics.

It was obvious that spatial magnitudes and those connected with motion could be expressed mathematically. Geometry deals with space and calculus with change and motion. But there exists a philosophical category of quality referring to these aspects of the world which can be grasped by our senses but which seem to be too rich and complex to be put into mathematical formulae. However, as time went on even the category of quality surrendered to mathematics. Already in Newton's *Optics* the first signs appeared that colors could be described mathematically, and when the wave theory of light entered the scene, the problem was settled. It was Newton who elaborated the mathematical theory of wave motion, and now this theory had only to be adapted to the new situation. Color corresponds to the length of a wave. The medium in which light waves propagate was called ether, and for the time being this solution seemed to be satisfactory.

The question of sound was even easier. Sound waves propagate in the air. It was not difficult to describe mathematically the wave motion of the air particles. As soon as this had been achieved, acoustics made quick progress.

At the end of the 17th century Locke attempted to clarify the situation. He distinguished primary qualities and secondary qualities. Primary qualities, such as extension, impenetrability and mobility, are the real properties of bodies. Secondary qualities, such as colors and sounds, are but the reactions of our senses to external stimuli. Secondary qualities belong to the field of physics which investigates the stimuli that are causing them and also to the field of psychology which investigates human reactions to these stimuli.

Are the properties of being cold or warm primary or secondary qualities? Bodies are doubtlessly cold or warm in a different way than they are green or red. Heat, contrary to color or shape, can "flow" from one body to another body. The theory of a "heat fluid," which can flow (like an ordinary fluid) from a warmer body to a colder body, was a compromise between the understanding of heat as a quality that can only be felt and the understanding of heat as a quality that can also be measured. The classical thermodynamics that entered the scene in the 19th century mercilessly eliminated the idea of a "heat fluid" and reduced the concept of heat to that of the kinetic energy of the particles constituting a given body. Heat is thus connected with the motion of particles, and can be measured just as motion can be measured. This was a typical reduction of a quality to a quantity, and also a great success of the mechanistic world view, claiming that everything can be reduced to the mechanical motions of material particles.

Obviously, it is impossible to trace the motions of all individual particles constituting a sample of a gas or fluid, but one can investigate such motions with the help of statistical methods. When in the second half of the 19th century it turned out that thermodynamics is only statistical mechanics, it became clear to the great majority of physicists that to show that all of physics is but an "applied mechanics" was only a question of time and calculation skill.

THE PEOPLE WITH LONG EYES –
SCIENCE SEES MORE THAN OUR EYES DO

When astronomers first began to develop the world's largest observatories in Arizona, USA, the indigenous Americans named these scientists "The People with Long Eyes", a primitive expression in the richest sense of primitive. The natives "saw" what was really happening. We humans were beginning to extend our curiosity about the universe with new technology; but it was more than technology. It was an adventure into the wonders of creation. The night sky is wonderful; it draws us to marvel. We humans respond to its call in many different ways: poetry, art, literature, children's fairy tales, etc. But science sees beyond what the human eye can see. It does this not just with the help of sophisticated optics and modern electronics which enlarge the natural power of our eyes but also, using mathematical models, it can dissolve the stream of photons, the minuscule quanta of light, into meaningful and wonderful, all-inspiring information.

All images were taken by the Vatican Advanced Technology Telescope at the Mount Graham International Observatory, Arizona.

Fig. 1 [UGC 12343] Here we see a beautiful galaxy with the catalogue label UGC 12343. This image, made with a near-ultraviolet filter in November 1996, was presented to Pope John Paul II at a meeting of the Pontifical Academy of Sciences on the formation of galaxies. It is a typical barred galaxy situated in the Pegasus constellation. Along the bar we can see luminous condensations, most probably the locations of intensive star birth processes. The galaxy is 105 million light years away and is participating in the expansion of the universe: its recessional velocity is 2381 km/s. On 27th July 1990 an explosion of supernova was observed in it

Fig. 2 [NGC 6781] During the lifetime of a star, equilibrium is maintained between the radiation pressure acting outwards and the force of gravity acting inwards. When the supply of nuclear fuel in the star is nearly exhausted, the equilibrium is perturbed and the dying star first oscillates and finally explodes. If the explosion is very powerful, the phenomenon is called a supernova. Here we see a bubble of gas that has been emitted by a dying star. The object, NGC 6781, is situated in the Eagle constellation. It is 2500 light years away. The bubble is approximately two light years across and is continuously expanding. Such processes play an important role in the life of a galaxy since they return to interstellar space the stellar material enriched in heavy elements

Fig. 3 [NGC 3147] Here we see spiral galaxy NGC 3147 in the constellation Draco showing tightly wound spiral arms around a clearly visible nucleus. This galaxy belongs to the Seyfert class of galaxies: their nuclei produce a strong outflow of highly ionized gas, visible as emission lines in the spectra. Karl Seyfert was an astronomer who, in 1943, first identified this class of galaxies. The nucleus of NGC 3147 hosts a massive black hole. It can be regarded as truly an edge of space-time. The gravitational field surrounding the black hole is so strong that even quanta of light that cross the boundary (justly called a no-return surface) are trapped inside with no possibility of escaping

5145 Pholus

Fig. 4 [5145 Pholus] And now, something from our closer astronomical neighbor-hood. In May 2003, asteroid or comet nucleus Pholus (5145) happened to be passing by the galaxy NGC 5964. Astronomers took three separate images, in red, green, and blue, and then combined them to make this image; but since Pholus was moving between one image and another, it appears here as a rainbow "streak"

3. THE IDEA OF FIELD

In the 19th century the phenomena of electricity and magnetism started to attract the attention of physicists. Could these phenomena also be reduced to purely mechanical interactions? Newton convinced his successors that in the mechanistic view of the world there could exist forces acting at large distances. Their prototype was gravity. Therefore, it did not come as a surprise that an electric charge acts on conductors even if it does not touch them. A similar property of magnets was known from antiquity. However, when more was learned about the geometry related to action at a distance (for instance, by studying the force lines along which iron particles align in the neighborhood of a magnet), the idea of electric and magnetic fields seemed natural and appealing. The space surrounding the electric or magnetic charges acquired more and more physical features.

Mathematically a very beautiful theory unifying electricity and magnetism into one electromagnetic interaction was proposed by a Scottish physicist, James Clark Maxwell. Although he used mechanical models to help his imagination, he rejected them totally when the theory was ready. The electromagnetic field equations turned out to be an abstract mathematical structure independent of any mechanical images. One can read out of these equations that any electromagnetic perturbation spreads in the form of a wave, but all attempts to put into the picture a medium in which this perturbation propagates were no more than a concession to our imaginations that were unable to cope with the new idea.

The difficulties increased, creating a crisis in classical physics. The crisis soon became a catastrophe that lead to the one of the greatest revolutions in the very foundations of physics. It is not infrequent that science elaborates concepts that become the property of the whole of human culture. Newtonian mechanics gave our culture a handful of concepts without which, as we are inclined to think, our culture would not have been able to function. Our everyday understanding of time, space, causality, and matter strongly depends on what classical physics had to say on these notions. Typical examples of other such concepts are the concepts of evolution and field. The former entered into "social circulation" through the Darwinian theory of natural selection, and the latter through Maxwell's electromagnetic theory. The concept of evolution has had a great career, becoming almost a slogan good for every occasion. The concept of field cannot be compared to that, but it also went far beyond the domain of applicability in physics and is present in many philosophical analyses.

In mathematics and physics the field concept is now an irreplaceable tool of inquiry. When in the 20th century new physical interactions were discovered (weak and strong nuclear interactions) there was no doubt that they were of a field character, and when Einstein began thinking about the unification of all physical forces, he spoke of the unified field theory. No wonder, therefore, that mathematics (first, for its physical applications and then for its own purposes) amply elaborated the so-called field theoretical methods. The idea of vector field, initiated by Maxwell himself, gave rise to various fields considered in contemporary mathematics (vectorial, tensorial, spinorial, and other fields). Today, one can hardly imagine modern geometry without the techniques of computation related to various fields. This could be regarded as the scientific counterpart of a philosophical standpoint that there is no absolutely empty space but rather that every space is a carrier of various fields.

4. RELATIVISTIC REVOLUTION

The great revolution in space and time physics began at the end of the 19th century with the discovery that Maxwell's electrodynamical theory could not be put into the framework of classical physics. This was by no means an easy discovery. It was preceded by many attempts to reduce the "electrodynamics of moving bodies" to the Newtonian physics of motion. However, these attempts were not only unable to cope with deep theoretical problems, but also led to inconsistent empirical results.

The way out of this critical situation was found by Albert Einstein. His special theory of relativity was another success of the mathematization process of space and time. Classical physics had already introduced some counter-intuitive elements into the body of physical knowledge about space, time and motion, but our imagination so readily accommodated them that today we call them common sense. The domain of high velocities (comparable to that of light) remains entirely beyond the reach of our senses, and the only thing our imagination can do is to claim that in these domains everything functions in exactly the same manner as in the world that surrounds us. It was the method of mathematical modeling of the world that overcame the limits of our imagination and opened the domain of high velocities to scientific exploration. Einstein had taken seriously the results of the experiments that so deeply troubled his predecessors, and he based them on the formulation of starting assumptions. All the rest was the result of mathematical deduction. The consequences of this strategy, although counter-intuitive, were corroborated by many new experiments. Physical reality surrenders to mathematical models rather than to our imagination.

Both physicists and historians of physics agree that the special theory of relativity was, in a sense, the result of the "inner logic" of scientific development, i.e. if not Einstein in 1905 then somebody else would have formulated it soon afterwards. But, on the other hand, the general theory of relativity constituted the outcome of Einstein's solitary struggle with a problem that nobody before him had even noticed. Einstein's profound mathematical intuition together with his long mathematical inquiries (Einstein had also to study new chapters of mathematics) did the job. The road from initial assumptions to the final shape of the general theory of relativity was much longer and more tedious than in the case of the special theory of relativity. Had Einstein not guessed the correct mathematical structure (the gravitational field equations), physics would have been poorer by one of its most elegant theories. Later on Einstein was accustomed to say that there are two criteria of the truthfulness of a physical theory: first, agreement with experimental results and, second, its inner perfection. In 1915 when he formulated his gravitational field equations, he had at his disposal only three empirical tests for his new theory. From among all of them only the perihelion motion of Mercury provided the concrete number that could be compared with the theoretical prediction. But even this number, representing a small deviation from the prediction following from Newton's theory, was not determined very precisely. Only much later did the two remaining tests become precise enough to be used in practice. Consequently, Einstein's first criterion (agreement with experiment) was not of great help. Einstein was then sentenced to use his second criterion (inner perfection). And he used it amply. To be sure, it is a subjective criterion. But only to some extent. The history of mathematical and physical discoveries suggests that there are some people who are able, in a fairly objective way, to distinguish a beautiful mathematical structure from many other mathematical structures that are lacking in such beauty.

In the mathematical structure that was used by Einstein to model gravity, there is encoded an enormous quantity of information about the world. In subsequent decades many solutions of Einstein's equations were found, and still continue to be found. Some of them model phenomena the existence of which Einstein was unable even to suspect. This was the case with solutions representing models of the universe and models of massive stars. In some cases, the equations turned out to be "wiser" than Einstein himself. For instance, Einstein stubbornly resisted accepting the fact that the universe expands, in spite of the fact that the equations clearly indicated this. He even modified the equations, adding to them the famous cosmological constant, to obtain the static (non-expanding) solution. Some ten years later astronomers discovered that galaxies recede from each other with velocities increasing with their distances from us. The universe really expands. Einstein had to acknowledge his defeat.

Also in the case of massive stars there was a surprise. Soon after Einstein had written his field equations Karl Schwarzschild (in the same year, 1915) found a solution modeling a very strange configuration of matter. Only about half a century later did it turn out that such configurations really do exist. They are outcomes of late stages of the evolution of very massive stars and are now called black holes. How could the equations know something that would be discovered fifty years after they had been written down? And this is exactly what often happens. Even today we find new solutions of Einstein's equations that model various things of which Einstein could have had no idea. Some of these new solutions represent exotic worlds, which at first glance seem to have nothing in common with our world but which soon after reveal some unexpected properties of our cosmos. Some other solutions represent objects which are later found observationally (e.g. gravitational waves); some others turn out to be useful in our investigations of the very early epoch of cosmic evolution (e.g. inflationary models or cosmic strings).

The general theory of relativity, together with its applications to cosmology and astrophysics, could be regarded as the greatest achievement of the mathematical-empirical method, if not for the fact that it has, in this respect, a very strong competitor, namely quantum mechanics, the theory which, with the help of experiments and mathematical models, penetrates the world of elementary particles and the most fundamental physical interactions.

5. QUANTUM REVOLUTION

The road to quantum mechanics was even more tedious than that to general relativity. Quantum mechanics, unlike general relativity, is the fruit of the collective work of many researchers, both experimentalists and theoreticians. In the beginning some experimental results and theoretical data indicated that classical physics, when applied to the atomic scale (the smallest scale known at that time), led to paradoxes, inconsistencies, or even contradictions. The necessity to remove, or at least to relax, these unwanted effects forced upon physicists more and more deviations from known classical methods and laws. This way of proceeding produced a conglomerate of heuristic rules and working models (which were later collectively called an older quantum mechanics). This process was accompanied by heated discussions and interpretative battles caused by the fact that the image of the microworld that slowly emerged out of these attempts was very different from the common sense standards established by classical physics. The elements of this great puzzle started to yield a coherent whole only when Paul Dirac discovered a mathematical structure that unified the two different formulations of quantum mechanics proposed previously by Erwin Schrödinger and Werner Heisenberg.

The mathematical structure discovered by Dirac is now called the theory of Hilbert spaces and of the linear operators acting on them. Any Hilbert space is a vector space equipped with some additional properties that make out of it a very rich and elegant mathematical entity. Owing to these properties Hilbert spaces are no less and no more rich than is required by quantum phenomena. Every vector in a Hilbert space represents a state of a quantum system (e.g. a state of an atom or an electron).[3] A linear operator on a Hilbert space is a mathematical object that transforms one vector in this Hilbert space into another vector in this Hilbert space or, in physical language, a mathematical object describing a process in which a considered quantum object (an electron, say) changes its state. This can correspond to the measurement process in which a quantum system changes its state because of its interaction with the measuring device. In classical mechanics the act of measurement does not disturb the measured object (measuring my car's velocity does not significantly change this velocity); this is no longer true when quantum objects are concerned (for instance, the state of an electron changes under the impact of the measuring process). By suitably choosing linear operators that correspond to experiments really performed we gain an insight into the world of atoms, elementary particles and quantum fields, the world that is completely inaccessible to our senses.

The effectiveness of this method relies on two facts. The first is the existence of a mathematical structure (linear operators on Hilbert spaces) which in some mysterious way reflects the structure of the quantum world. The second is the existence of sufficient quantum effects that can be strengthened so as to leave measurable traces in macroscopic measuring devices. Measurements really performed translated into the language of linear operators acting on a suitable Hilbert space allow us to establish a correspondence between the structure of this Hilbert space and the structure of the quantum world. Owing to this we can read off the structure of the quantum world from the structure of Hilbert spaces and suitable operators acting on them.

Taking into account the enormous achievements of quantum mechanics and of modern quantum field theories, and the fact these theories form the very fundaments of contemporary physics, we are entitled to regard them as the most successful fruits of the mathematical-empirical method.

6. THE MATHEMATIZATION OF CHAOS

Does there exist a domain of physical phenomena that cannot be mathematized? Until quite recently, it was thought that such a domain consists of chaotic processes. How could one put into mathematical formulae what happens at the bottom of the Niagara Falls? Or the formation and sudden disappearance of bubbles on the surface of boiling water? Or the shapes of trees or of mountain ridges? However, it has quite recently turned out that behind all these phenomena there is a hidden order. Although there are no two identical trees in the world, everyone can easily distinguish an oak from a poplar. If every bubble on the surface of boiling water consists of millions of particles, there must exist something that tells all these particles how to coordinate their movements so as to form a coherent bubble. And although water particles in the Niagara Falls form instantly changing configurations, when one looks at the Falls from a certain distance, it always presents the same view.

It is the theory of chaos, also called the theory of deterministic chaos, that explains all these phenomena. This theory was initiated by Poincaré at the beginning of the 20th century, but it remained almost unchanged during the next fifty or so years. Only recently, its development has undergone rapid acceleration. It has turned out that, contrary to previous estimates, no highly complex mathematics is required to cope with chaotic phenomena. In fact, dynamical equations modeling such phenomena could even be very simple. They must exhibit only one property, namely a high sensitivity to changes in their initial conditions. All bubbles on the surface of boiling water are alike since they are governed by the same system of differential equations (such a system is called a dynamical system), but some bubbles are bigger than others, some live longer than others, etc., because small changes in the initial conditions of these equations lead to very different solutions.

The theory of chaotic dynamical systems has caused a great revolution in physics in the domain hitherto believed to have been a closed chapter, namely in classical mechanics. The discovery that the great majority of classical dynamical systems are chaotic was a real surprise. Indeed, in mathematical parlance, almost all classical dynamical systems are chaotic, i.e. all classical dynamical systems are chaotic with the exception of perhaps a finite number of them.

These achievements have thrown new light on classical determinism. Of course, equations of motion of classical mechanics are deterministic in the sense that by knowing their initial conditions with unlimited preciseness one could uniquely compute the past and the future behavior of the system. However, in practice any measurement can only be done with a finite precision (even if we disregard quantum indeterminacies), and consequently the initial conditions of any equation cannot be known with unlimited preciseness. In such a case, we are dealing with a chaotic dynamical system, the future behavior of which is unpredictable.

As we can see, the mathematization of physical phenomena does not entail their strict predictability. There are physical phenomena which can be fully treated by mathematics which, however, have an "open future." The power of mathematics and its application to natural phenomena are much greater than it was possible to suppose.

7. THE MATHEMATIZATION OF "EVERYTHING"

Encouraged by the enormous successes of mathematical physics, physicists dream their dream of the final theory of everything.[4] The idea of unification was, from the very beginning, present in modern physics. It is enough to remember that modern physics was born together with the unification of earthly physics with the physics of the heavens: the same law of gravity governs the fall of an apple and the motions of distant stars. Soon after, physics divided into many branches, each quickly developing in its own manner, but all using essentially the same methods, and all participating in the process of creating the same body of knowledge and contributing to the understanding of the physical world. Moreover, as various branches of physics became more theoretical and more mathematical, the borders between them became more and more fuzzy. In the 19th century, it turned out that thermodynamics was but a form of statistical mechanics, and Maxwell's electrodynamics unified two domains separate until then: that of electric phenomena and that of magnetic phenomena.

It was Einstein who first clearly formulated the unification program of physics. After creating his general theory of relativity, in which the gravitational field was represented as a geometric deformation of space-time, he understood that electromagnetic phenomena should also be included in this geometric framework. Many attempts to implement this task, undertaken by Einstein himself, Weyl, Kaluza, Eddington and others, failed to produce the desired results. However, these were not completely in vain: owing to them many elegant mathematical structures were discovered, and many new ideas were injected into the stream of physical knowledge. For instance, Hermann Weyl took an essential step toward the formulation of gauge theories which now serve as mathematical models of all physical fields except gravitation (there are also attempts to include gravity into the gauge scheme). Kaluza and Klein considered space-time with an additional dimension that was supposed to "make room" for electromagnetism in Einstein's general relativity. Today, multidimensional spaces are an indispensable tool in some recent unification attempts such as superstring theory.

Nowadays we know that the original Einstein unification program could not succeed. Einstein was not aware of the fact that besides electromagnetic and gravitational forces there exist also nuclear strong and nuclear weak forces, and that they must also be included in the unification scheme. The gauge method discovered by Weyl turned out to be very useful in this respect. With the help of this method, suitably adapted, Steve Weinberg and Abdus Salam accomplished, in the 1970s, the unification of the electromagnetic force and the weak nuclear force into one force, now called the electroweak force. This unification has been empirically verified in the great particle accelerator at CERN near Geneva. According to the Weinberg-Salam theory, electromagnetic and nuclear weak forces are independent of each other when the environment temperature is less than about 100 GeV (Gigaelectron-volt), and above this temperature they become one electroweak force. When at CERN such temperatures became available, the empirical verification of the electroweak unification was accomplished.

There are good reasons to believe that at temperatures of the order of 10^{14} GeV, the electromagnetic force unifies with the strong nuclear force. This is often called the Grand Unification of physics. There exist several scenarios of this unification, the more important of them being based on the gauge method, but to test them empirically is a difficult task, since temperatures of that order will not be available in the foreseeable future. The only more or less realistic possibility is to test them indirectly with the help of cosmology. Temperatures of the order of 10^{14} GeV dominated the universe some 10^{-35} sec after the Big Bang, and if we put the Grand Unification mechanism into a cosmological model, we can compute some traces these mechanisms had to leave at later epochs of cosmic evolution. It is, in principle, possible to detect such traces with the help of astronomical observations. If this were the case, it would be possible indirectly to test Grand Unification models. However, so far this method has produced no results.

The same remarks, but with even more caveats, apply to the so-called Super-unification of Physics, i.e. theories trying to unify all physical forces, gravity included. The theory of superstrings, together with its newest ingenious generalization, the so-called M-theory, is the best known attempt of this kind. Superstring theory is a combination of two ideas: the string-idea and the "super-idea." The string idea consists in the assumption that elementary particles are not point-like objects but rather string-like objects. The great success of this approach is the demonstration that various vibration models of these one-dimensional strings give the correct spectrum of known particles (plus many not yet discovered), and that interactions between them can be reduced to a geometric behavior of strings. The "super-idea" goes back to an older theory of supersymmetry. There are two kinds of elementary particles: fermions and bosons. The constituents of ordinary matter (such as protons, neutrons, electrons) are called fermions; particles that transfer interactions between other particles are called bosons (e.g. a photon is a boson; it transfers electromagnetic interaction between electrons).

Supersymmetry is an operation that transforms fermions into bosons, and *vice versa*. Mathematically, it is a very elegant operation. It involves, among others, replacing real and complex numbers by the so-called Grassmann numbers (called also Grassmannians). By replacing ordinary numbers by Grassmann numbers one can construct many new mathematical structures, for instance: supervector spaces, supergroups, supermanifolds, and so on. However, we should emphasize that this replacement is by no means a simple procedure. It often requires a great deal of ingenuity.

Although it is not possible to directly check Grand Unification and Super-unification theories, the program of physics unification has brought a very interesting result – a unification of physics and cosmology. As we have mentioned above, the temperatures needed for Grand Unification were reigning in the universe when it was 10^{-35} sec old. One can readily compute that the temperatures required for Superunification, which are of the order of 10^{19} GeV, were reigning in the universe when it was 10^{-44} old. And again, by inserting Superunification theories into a cosmological model we could try to compute and then to discover their traces at later epochs. What cannot be done with the help of our particle accelerators, can be done with the help of cosmology. In this sense, physicists sometimes say that the universe is an "Ultimate Particle Accelerator."

8. THE PLANCK THRESHOLD

Superstring theory and M-theory go even further and try to reconstruct the pre-Planck era. These investigations are really exciting, but at the moment they find themselves at the very limit of the scientific method. This book, the main goal of which is to explore the essence and effectiveness of this method, should stop here. Within the secure limits of the method the following picture emerges of the deepest known structure of the universe. In the beginning there was an extremely rich but, from the geometric point of view, very simple mathematical symmetry: we can call it the Primordial Symmetry or Primordial Supersymmetry (not necessarily having anything in common with the supersymmetry known from the present supergravity or superstring theories). Subsequent violations of this Primordial Supersymmetry (decouplings of the presently known physical forces) led to the growth of more and more diversified multiplicity. The physicist's dream of the ultimate theory is but a dream about the Primordial Symmetry from which all the present richness of the world originated.

This does not mean, however, that the whole of the future history of the cosmos, up to its smallest details, was once and for all predetermined in the Primordial Symmetry like a melody is once for all predetermined on a magnetic tape or a CD. Subsequent symmetry breakings and processes, leading to the emergence of more and more complex structures, have introduced into the history of the world's evolution elements of unpredictability. Moreover, there are strong reasons to think that the most fundamental level of physics is of a quantum character and, consequently, that quantum indeterminacies are built into the most fundamental laws governing the structure of the universe. And precisely these laws, which are ultimately responsible for the subsequent symmetry breakings, after being suitably averaged, lead to the classical determinism of our macroscopic world. However, as we have seen above, even on this level the world abounds in a great variety of chaotic processes. They are unpredictable and often generate new complex structures. The shape of our planet, together with its biological systems in which we are participants, are products of these creative forces.

The evolution of the universe is mathematical in its character. But the mathematics which lies at the basis of this evolution is by no means dull. It is not like always solving the same computational exercise, but rather like continuously posing a new problem that must be solved.

9. LIMITS OF THE METHOD

At the end of our inquiries the question is unavoidable: can everything be mathematized? The question remains valid even if by "everything" we mean *everything* physics is dealing with. If we ask the question of the limits of scientific method, Gödel's theorems immediately come to mind. If in pure mathematics there are such strong limitations, what should be said about physics?

When David Hilbert was working on the formalization program of the whole of mathematics, he expressed the views shared by many mathematicians. As is well known, the idea of an axiomatic system is regarded as the highest standard of formalization. A certain domain of mathematics becomes an axiomatic system if one succeeds in finding such a set of assumptions (called axioms) and such a set of inference rules that all mathematical truths (all theorems) of this domain can be inferred (deduced or proved) from these assumptions with the help of these inference rules. Inferring theorems from axioms, within a given axiomatic system, is almost a mechanical procedure. The "human factor", which is often a source of many inaccuracies and errors, is here reduced to its minimum. As a matter of fact, Hilbert's program consisted in reducing the whole of mathematics to a single overwhelming axiomatic system which would certify many "smaller" axiomatic subsystems. This program was not only the result of the fact that mathematicians like preciseness and economy of thinking; it was supposed to be, first of all, a remedy for the problems and paradoxes which had been invading mathematics for some decades.

In such a situation, the Gödel theorem appeared. Gödel proved that if an axiomatic system, at least as rich as arithmetic, is complete then it is self-contradictory. Let us explain the concepts. An axiomatic system is complete if all true propositions of a given domain can be inferred from its axioms. An axiomatic system is self-contradictory if from its axioms can be deduced both a certain theorem and its negation. In fact contradiction ruins the very idea of an axiomatic system. Within a self-contradictory system (i.e. a system containing a contradiction) one can prove nothing. Gödel's theorem does not prevent constructing an axiomatic system containing the whole of mathematics, but it asserts that such a system, containing arithmetic, must be self-contradictory, and therefore useless. Gödel's theorem allows one to construct an axiomatic system free of contradictions and richer than arithmetic, but then there will be in it statements true in the domain to which the system is referring which, however, cannot be deduced from its axioms. In other words, Gödel's theorem says that there is no set of axioms and rules of inference strong enough to prove all true statements of arithmetic which, at the same time, would not be strong enough to "prove" also false statements. If this theorem is valid with respect to arithmetic, it is also valid (even "more valid") with respect to the whole of mathematics.

Hilbert's program cannot be implemented. Mathematics cannot be presented as a single axiomatic system. When a physicist applies mathematics to investigating the world, he adds to the limitations of mathematics additional limitations that follow from the fact that he applies mathematical structures to something that is not mathematics. Therefore, we should expect that the mathematization of physics will meet two kinds of limitations: the one inherent in mathematics itself, and the other that is associated with the application of mathematics to physics.

It seems that there is only one narrow gap through which we could bypass Gödel-like limitations in physical research. As we have mentioned, Gödel's theorem remains valid only for axiomatic systems that are at least as rich as arithmetic. Would it be possible to reduce the whole of the mathematics used in physics to but a part of arithmetic? Everyone who has had some acquaintance with theoretical physics is instantly inclined to give a negative answer to this question. However, perhaps the world is like a computer which uses only the simplest numerical operations, and only we, who are lacking the computer's abilities, must employ functional analysis, differential geometry and all the rest of highly complicated mathematics in our investigations of the world? The suggestion contained in this question seems to be very unlikely, but it cannot be *a priori* excluded. There are quite strong reasons to believe that nonlinear equations, deterministic chaos and many other mathematical structures that are essentially richer than arithmetic must be used in physics, not only as tools making our mathematical reasoning easier, but as essential elements of the structure of the universe. If the world is not only a "quick abacus" but a "physically interpreted" rich mathematical structure, then Gödel-like limitations must somehow be impressed on the world's structure.

For the time being, however, mathematicians experience much more discomfort than physicists because of the "shaky foundations" of their discipline. Moreover, in looking for a firm ground for progress, mathematicians often turn for help to physicists. If the applications of mathematics to physics give such wonderful results, the state of mathematics cannot be that bad.

10. A FIELD OF RATIONALITY

Let us once more take a look at Hilbert's program. In his unfulfilled vision, the whole of mathematics is a single, overwhelming axiomatic system consisting of a great number of "smaller" axiomatic subsystems, each of them being an axiomatization of a particular branch of mathematics. All these subsystems are interconnected by manifold logical dependencies. In this superstructure of structures everything is strictly formalized, leaving no space for inaccuracies and antinomies, and it gives us the certainty, once and forever, that mathematics is free of any contradiction. Mathematicians had been so struck by the transparency of Hilbert's program that when it became clear, owing to Gödel's theorem, that it could not be implemented, they were deeply distressed by their unfulfilled dream of mathematical unity.

And what if mathematics is something much greater than a single axiomatic system? Let us imagine a set of all possible mathematical structures. When we think about a huge agglomeration of objects, the term "set" comes first to mind, although in the case of "all possible mathematical structures," the term "set" almost certainly is not applicable in its standard meaning in set theory. Let us then use the term "field." This term seems to be more suitable since, when thinking of mathematics, we should not think about a loose agglomeration of objects, but rather about a network of manifold structures connected with each other via various inference relations. Such a "field" is, in a sense, a field of potentialities. It contains not only mathematical structures known to us, but also structures which will be discovered in the future, and even structures which never will be discovered but are, in some sense, possible. To emphasize that this field contains not only all possible mathematical structures, but also all possible logical dependencies among them, the name "formal field" could be employed. For obvious reasons, the name "field of rationality" is also sometimes used.[5]

The field of rationality is clearly not an axiomatic system. It has no distinguishing parts or domains which would play the role of "universal axioms" or of "universal rules of inference." Its potentiality prevents it from qualifying as an axiomatic system, and underlines its unlimited possibilities. At the present stage the idea of this field is vague and imprecise, but we can make use of it as a not-yet-ready tool to grasp the philosophical meaning of Gödel's theorem.

Let us suppose that we are constructing an axiomatic system (e.g. a system of arithmetic). When we are formulating our axioms, we pick up a certain point, or a domain, within the formal field. Starting from this domain, with the help of a chain of deductions we can go to another domain of this field. Going, in this way, in various directions we can create a network of deductive chains within the formal field. If we remain in a restricted domain of this field, everything goes well, and we can construct a good axiomatic system. However, if this network becomes too extended, contradictions or other logical pathologies can appear between deductive chains "crossing each other." In other words, an axiomatic system, rigorously understood, cuts off from the formal field some of its domain. If we want to protect this domain from contradictions, it must be sufficiently small (less rich than the system of arithmetic). But in such a case, our axiomatic system turns out to be incomplete. Some of its true theorems are situated beyond the deductive reach of the axioms. Of course, these theorems can be added to our axiomatic system (e.g. as new axioms) but then the domain controlled by our axiomatic system becomes larger, and contradictions are imminent.

The rationality field concept can be regarded as a tool for interpreting Gödel's theorem. The main idea of this interpretation suggests that mathematics is a holistic system. If one wants to isolate from the whole a rich enough domain of the field, inconsistencies appear leading either to incompleteness or to contradiction. Gödel's theorem establishes the conditions under which this happens.

The rationality field concept could also be interpreted in a more ontological manner as something that exists in a certain sense, and that conditions both the possibility of mathematical structures and their effectiveness in modeling the real world. If this ontological interpretation is correct, the world exists because it remains in a very intimate relationship with this field. In this context the name "field of rationality" is especially fitting. The fact that the universe can be rationally investigated finds its justification in this field.

AFTERTHOUGHTS

~

1. INTRODUCTION

The decisive step in the evolution of rationality was the discovery of the scientific method. In spite of the fact that it was anticipated by a long (and laborious) historical process, its appearance in the 17th century can be compared with the mutation that initiated Greek philosophical thinking. The birth and consolidation of the scientific method created a new type of rationality. This type of rationality is undoubtedly rooted in the tacit assumption that the world possesses a property owing to which it can be rationally investigated, but the world does not respond equally well to all of the questions that are addressed to it. As long as people were asking questions based on purely conceptual analysis, progress was hardly noticeable; however, as soon as people started to use mathematical models and to verify them with the help of observation and experiment, progress became rapid. This allows us to claim that the rationality of the world is of a mathematical type.

Progress in the sciences has been spectacular: from classical physics through to quantum mechanics and general relativity with their far-reaching applications, quantum field theories and relativistic cosmology, until the present search for the superunification of physical interactions. Although it is always the same mathematical-empirical method, we can see it clearly in some developmental trends. At the early stages of classical physics it seemed obvious that physics consisted of empirical data and mathematical theories; experimental data are needed to provide a basis for a given theory, and then to check its predictions. Various theories covered various domains of physics. As progress went on, more and more theories were unified, and the borderline between empirical data and theories became more and more fuzzy. It is now clear that there are no "naked" empirical data; every empirical datum is "theory-laden". In the more advanced theories of modern physics this "impregnation" of empirical data with theory is indeed far-reaching. For instance, in a modern particle accelerator the results of experiments consist in long chains of zeros and ones in the outputs of giant computers. Without highly mathematical theories it would not be possible either to obtain these results, or to find their correct interpretation. Even to construct a particle accelerator a huge amount of theoretical work is required. The accelerator itself can be regarded as a big "theoretical factory" (a kind of software) incorporated into a technological system (a kind of hardware).

Some philosophers of science complain that in this way the empirical character of physics is gradually lost. On the contrary, we think that, precisely by this process, physics becomes more and more empirical. The mathematical-empirical method loses its dualistic character (experiments and theory) and becomes "monistically more coherent": experiments and theories are but poles of the same way of formulating questions directed to the universe and enforcing on it at least some answers.

It is often said that the greatest discovery of science is its method. We agree with that. However, the scientific method is not only a collection of research tools and instructions for their use. In the method itself, which has proved so effective, there is contained some information about the world. The world must have some property, or a set of properties, owing to which exactly this method, and not some other one, works so well. In fact, this idea underlies all of our considerations in the present book. As the book comes to an end we wish to make this idea even more evident. When dealing with a Big Question, it is a good thing to go back to Old Masters who, not yet immersed in too many technicalities, were able to see deeper and further. Let us then go to Leibniz.

2. GOD'S MATHEMATICS

Gottfried Wilhelm Leibniz appeared in our book only occasionally, mainly when we were speaking about the origin of calculus. This is a serious omission. Leibniz is not only one of the most original and most productive philosophers, he is also the first thinker who regarded the existence of modern mathematical physics as a subject worthy of deep philosophical analysis. His aim, in this respect, was exactly the same as that of the present authors.

In 1677 Leibniz wrote an essay entitled *Dialogus*. In the margin of the text we find a short sentence in Latin written by him: *Cum Deus calculat et cogitationem exercet, fit mundus*, which, in English translation reads: "When God calculates and thinks things through, the world is made". As often happens when profound ideas are compressed into a short sequence of words, one has to invest a lot of time and effort to decipher the true meaning.

Everybody has some experience of dealing with numbers. When one is confronted with not very large numbers, the calculation is almost a routine process, and if one has mastered basic mathematical techniques, the same can be said for dealing with big numbers. True mathematical thinking begins only when one has to solve a more complicated problem, to formulate and then to prove a theorem or – to put it more explicitly – when one has to identify a certain mathematical structure, to understand the principles of its functioning, to construct a new structure starting from known ones, to grasp its intimate connections with other structures, etc. Such manipulation of structures is usually connected with calculations since mathematical structures like to dress up in numbers and algebraic formulae. The language of calculations is their natural language.

It is more or less such an image that we should connect with Leibniz's metaphor of a God who calculates. God is seen as calculating and thinking things through in His one act of creation. Things thought through by God, in this context, could best be identified with mathematical structures as patterns of the created world. To better grasp Leibniz's idea, we could imagine his work when he was creating the differential and integral calculus. First, he had to identify the problem and to collect partial elements of its solution spread throughout the writings of his predecessors. Then a few decisive generalizations had to be made, new theorems formulated, expressing dependencies between some aspects of an emerging structure, and some examples computed to check whether the newly born structure worked correctly. If it did, a new mathematical world had been created.

To render Leibniz's metaphor more transparent, one must free it from all human limitations and add an important element: that for God to obtain the result is to instantiate it. When a mathematical physicist creates a new mathematical structure and tries to apply it to model physical situations, he or she defines concepts so as to fit them to empirical results. If necessary, the theoretical physicist modifies starting definitions, and accordingly synchronizes the entire network of relationships between various concepts. And when – after perhaps much trial and error – the mathematical structure reaches the adequate degree of maturity, the miracle of the method occurs. The mathematical structure becomes a physical theory that not only explains empirical data known up to the present, but is also ready to predict new data.

When God calculates and thinks structures through, all this human trial and error disappears. The world simply comes forth. In the following we shall try to put Leibniz's metaphor into the setting of the world that science studies.

3. MATHEMATICS AS A MORPHOLOGY OF STRUCTURES

It would be difficult to find the thinker who first used the term "structure" to denote the subject-matter of mathematical inquiry. Any in-depth study of mathematics, especially of such branches as geometry or abstract algebra, gives the impression that one is confronting an imposing edifice, all the elements of which fit perfectly together. It is, therefore, no wonder that a structuralist view of mathematics was current among working mathematicians long before it became a part of the philosophy of mathematics. It is the paper "Mathematics as a Science of Patterns" by Michael Resnik that is commonly regarded as the manifesto of the "structuralist movement"[1]. He claimed that the "objects" studied in mathematics, such as numbers or vectors, are unlike objects in the physical world, which are individual entities possessing "inner properties"; the "objects" of mathematics on the other hand are "structureless points" or "positions in structures". Outside of a structure, they are simply meaningless. For instance, number "three" acquires its meaning only if it is placed in the environment of other numbers. Even if we contemplate the number three alone, we know that it is greater than one and smaller than four, and that it consists of two units and one unit added together. If we erase these things from our mind, we erase the number "three". "Three" is but a place in the structure called real line.

When the idea that mathematics is a science of structures is understood in an intuitive manner, it does not excite any major emotions, but the agreement among philosophers of mathematics ends immediately as soon as one wants to give a technical definition of structure.

One very common approach to mathematics is based on the assumption that all of mathematics is built upon set theory. One could even speak of the set philosophy of mathematics. Since sets are typically regarded as objects, the supporters of mathematical structuralism would like to replace the set philosophy of mathematics by a structuralist philosophy of mathematics. These tendencies led people to look for help in category theory.

One would be tempted to say that structure is a network of relations. But relations presuppose objects between which these relations obtain, and we would like to get rid of objects. More precisely, relations are defined in terms of sets[2] and we are back in the set philosophy of mathematics.

Category theory is a mathematical theory with a strong structuralist flavor that has opened new perspectives. Very roughly speaking, by a *category* one understands: (1) a class of entities called *objects*, (2) a class of mappings between objects, called *morphisms*, and (3) compositions of *morphisms*. Of course, all these classes must satisfy suitable axioms. The important point is that objects need not be sets, nor need morphisms be mappings between sets. Although in the definition of category objects are mentioned, the stress is put on morphisms. There were even attempts to build an "objectless" category theory, but it did not lead to the anticipated results[3].

Examples of categories are: the category of sets, the category of topological spaces, and the category of groups. Objects of these categories are: sets, topological spaces, and groups, correspondingly; and morphisms are mappings between these objects, preserving their structural properties.

An important role in category theory is played by *functors*; they can be regarded as mappings between categories that preserve some of their properties. By investigating *functors* one discovers relationships between various mathematical theories (e.g. between the theory of sets and the theory of groups or topological spaces); in other words, one reconstructs some aspects of the structure of mathematics as a whole.

The non-initiated reader should not be troubled with the above technical terminology. What concerns us here is that things studied in mathematics can be grouped, with respect to some of their structural properties, into certain collections (categories), and relations between these collections (*functors*) establish dependencies between them by disclosing a way these collections are mapped into each other.

When Samuel Eilenberg and Saunders MacLane were creating the category theory, they aimed to grasp the essence of the structure idea, and to construct a tool to investigate dependencies between various mathematical theories. MacLane was also hoping that category theory could provide a better foundation for mathematics than that given by the theory of sets. When these hopes did not materialize he changed his views, and out of the failure he tried to create a new philosophy. He claimed that, although category theory does not provide the foundation for mathematics in the strict sense of the word, it has a "foundational significance", in the sense that it organizes the whole of mathematics into the "structure of structures". As expressed by E. Landry, mathematics is a science of "structures and their morphology"[4]. There have been attempts to express these rather rough ideas in a more precise way, and to define a "category of categories" that would formally implement the structuralist idea of mathematics. So far, however, the results of this highly technical program have been rather modest.

As we can see, the structuralist interpretation of mathematics is well-rooted in everyday mathematical research, but it turns out to be notoriously difficult to express it in a rigorous way. Moreover, we should admit that the borderline between "objectivist" and "structuralist" philosophies of mathematics is rather fuzzy: structures can be regarded as systems of relations between objects, or objects as "structureless points" in structures. However, all technicalities aside, we can safely agree, at least as a working hypothesis, that "mathematics is about structures and their morphology". Has this approach any implications for our understanding of the physical world?

4. STRUCTURAL REALISM

Surprisingly, the question asked at the end of the previous section is closely related to the recent return to the very old dispute in the philosophy of science between the defenders and opponents of the realist view of science. It is interesting that both defenders and opponents of this view look for arguments in the progress of science. Supporters of realism claim that "it would be a miracle, a coincidence on a near cosmic scale", if theories of modern physics such as classical mechanics, quantum mechanics or general relativity made so many very precise empirical predictions "without what they say about the structure of the world being correct, or at least approximately correct". On the other hand, anti-realists emphasize that successive revolutions in physics introduce striking discontinuities into scientific progress, and claim that the "scientific image of the world" presupposed by various scientific theories cannot be true if it is drastically changed by every major paradigm shift. The dispute was ignited by John Worrall who asked: "Is it possible to have the best of both worlds, to account for the empirical success of theoretical science without running foul of the historical facts about theory-change?"[5]. In his view, we should distinguish between continuity and discontinuity in the evolution of physics at the level of empirical results and at the level of the description of the world. There are no major problems with the accumulation of empirical results, and as far as the description of the world is concerned one should distinguish between intuitive images of the world and the mathematical structures employed by the physical theories on which these intuitive images are based. When a major physical revolution takes places "world images", indeed, often change in a discontinuous manner. For instance, the quantum image of the world is totally different from the classical image of the world. But there is a "smooth transition" from the mathematical equations of quantum physics to the mathematical equations of classical physics. Since mathematics is a science of structures, there is continuity between the structures of the world as they are described by the old physical theory and the new physical theory, although there is no continuity on the level of the images of the world presupposed by these theories. Therefore – concludes Worrall – the progress of physics does not provide evidence for "full-blown realism – but instead only for structural realism".

If mathematics is a "science of structures and their morphology", and if the essence of the method of physics consists in applying these structures to investigate the world, an unavoidable conclusion is that successive physical theories disclose, step by step, the structure of the world. And since this structure is a system of various substructures, one could say, analogously, that physics is a "science of the world structure and its morphology".

5. STRUCTURE OF THE WORLD

The next unavoidable question is: what should we understand by the structure of the world? The easy answer to this question would be that we should identify the structure of the world, as it is approximated by a given physical theory, with the mathematical structure this theory employs. But what if the physical theory in question admits of more than one mathematical formulation, as it is often the case in modern physics? For instance, quantum mechanics admits of several different mathematical formulations: in terms of operators on Hilbert spaces, in terms of Feynman's path-integrals, and in terms of density operators – to name only the most commonly known. This fact, indeed, creates a serious difficulty for a "naive realism", but it could be regarded as an asset for structuralist realism. The idea is to claim that, in the case of many different formulations of the same physical theory, we have various representations of the same structure. This is quite a common view among theoretical physicists. For instance, they easily agree that the wave formulation of quantum mechanics, created by Schrödinger, and the matrix formulation of quantum mechanics, created by Heisenberg, are but two different "pictures" of the same physical theory. James Ladyman seems to have correctly stated the solution: "The idea then is that we have various representations which may be transformed or translated into one another, and then we have an invariant state under such transformations which represents the objective state of affairs"[6].

The situation is not unlike that where the meaning of a book does not change when we go from one faithful translation of it to another (e.g. from its English to its German translation). Different translations are only different representations of the meaning of the same book. Translations are concrete entities that we can penetrate with our eyes and imagination. The meaning is an abstract entity that we can only reach by our power of understanding. To come back to Leibniz's metaphor, God calculates and thinks things through directly with abstract structures that we are able only to approximate, dimly and step by step, by laboriously studying their representations provided by our mathematical theories.

6. THE MIND OF GOD AND THE MIND OF MAN

Mathematics is a science of structures. Physics applies mathematics to study the world. Therefore the progress of physics discloses the structure of the world. If there is something more in the world than its structure, it is not open to the method of science. By using Leibniz's way of speaking, we could say that "God thinks things through", in terms of structures, and science, in its progressive evolution, gradually deciphers the Mind of God contained in His work of creation.

However, science is the product of the collective work of many human brains, but the human brain itself is a part of the world's structure; in fact, the most complex and the most sophisticated part of the world's structure. In the human brain, the world's structure has reached its focal point: the structure of the world has acquired the ability to reflect upon itself. This self-referential focal point is what we call the Human Mind. In this conceptual setting, science appears as a collective effort of the Human Mind to read the Mind of God from the question marks that surround us and of which we ourselves seem to be made. The Mind of Man and the Mind of God are strangely interwoven. This entanglement is a source and a driving force of science – the most adventurous adventure of humankind.

NOTES AND REFERENCES

～

CHAPTER 1

1 A. Einstein, "Physics and Reality" in: *Ideas and Opinions* (New York: Dell 1978) pp. 283–315.
2 Quoted by O. Pedersen, *The Two Books: Historical Notes on Some Interactions between Science and Theology* (Vatican City: Vatican Observatory Publications 2007) p. 25.
3 See Ref. [2] p. 6.
4 See Ref. [2] pp. 8–9.
5 See Ref. [2] p. 9.
6 A. N. Whitehead, *Science and the Modern World* (London: Harper Collins 1975) p. 14.
7 See Ref. [6].
8 *De Civitate Dei* VIII, 2. Air was a fundamental element in Anaximenes' philosophy.
9 *Stromata* V, 109.
10 *Timaeus* 53a.
11 "... the demiurge remained a strange and abstract deity to whom no altar was ever erected" O. Pedersen, *The Book of Nature* (Vatican City: Vatican Observatory Publications 1992) p. 13.
12 *Metaphysics*, Book XII, 1073a.
13 L. Kolakowski, *The Presence of Myth* (Chicago: The University of Chicago Press 2001).
14 K. P. Popper, *The Open Society and Its Enemies* (London: Routledge & Kegan Paul, vol. 2, 1974) p. 224.
15 See Ref. [14] pp. 230–231.
16 See Ref. [14] p. 232.
17 See Ref. [14] p. 231.

CHAPTER 2

1 This is the somewhat abbreviated story told by Plato in his *Republic*.
2 This shock can be heard in the name "irrational number."
3 *Timaeus*, 51E – 52B; translation by F. M. Cornford (London: Routledge and Kegan 1952).
4 See Ref. [3].
5 David Hilbert and S. Cohn-Vossen have written a beautiful book *Anschauliche Geometrie* (1932). It is worth reading the chapter dealing with "five Platonic solids." After reading it one can better understand Plato's fascination with the discovery of Theaetetus.

CHAPTER 3

1 *Metaphysics*, Book XIV, 1091a; translated by W.D. Ross, The Internet Classics Archive; http://classics.mit.edu
2 *Metaphysics*, Book XIV, 1093a.
3 *Metaphysics*, Book XIV, 1092b–1093b.
4 *Metaphysics*, Book XIV, 1092b.
5 *Metaphysics*, Book XIV, 1092b.
6 *Metaphysics*, Book I, 995a.
7 *Physics*, Book II, 193b; translated by R.P. Hardie and R.K. Gaye, Internet Classics, classics.mit.edu

CHAPTER 4

1 "The Sand-Reckoner", in: *Great Books of the Western World*, vol. 2, ed. by R. M. Hutchins (Chicago: University of Chicago Press 1978) pp. 520–526.
2 For instance, G. Sarton, *A History of Science. Hellenistic Science and Culture in the Last Three Centuries B.C.* (Cambridge MA: Harvard University Press 1959) p. 70.
3 O. Pedersen, *The Two Books: Historical Notes on Some Interactions between Science and Theology* (Vatican City: Vatican Observatory Publications 2007) pp. 12–17.
4 "The Method Treating of Mechanical Problems" See Ref. [1] pp. 569–570.
5 "Biographical Note. Archimedes, c. 287–212 B.C." See Ref. [1] p. 400.

CHAPTER 5

[1] A. Einstein, "Physics and Reality", in: *Ideas and Opinions* (New York: Dell 1978) pp. 283–315.

CHAPTER 7

[1] "On Prescription against Heretics", chap. 7; translated by P. Holmes, in: *The Ante-Nicene Fathers*, eds. A. Roberts, J. Donaldson, 10 volumes (New York: Charles Scribner's Sons 1896–1903).

[2] See Ref. [1] "On the Resurrection of the Flesh", chap. 3.

[3] See Ref. [1] "On Repentance", chap. 1; translated by S. Thelwall.

[4] D.C. Lindberg, "Science and the Early Church", in: *God and Nature*, ed. by D.C. Lindberg and R.L. Numbers (Berkeley – Los Angeles – London 1986) pp. 19–48.

[5] See Ref. [4] p. 24–25.

[6] See Ref. [4].

[7] *Epistula* 120, 3 (PL 33, 453); in Ref. [4] pp. 27–28.

[8] See Ref. [4].

[9] *Divine Institutiones*, 3, 24.

[10] Nicolaus Copernicus, "On the Revolutions of the Heavenly Sphres", Book 1, in: *Great Books of the Western World* (New York – London: Encyclopedia Britannica, Inc. 2007)

[11] "De Civitate Dei", XVI, 9 (PL 41, 487), in: O. Pedersen, *The Two Books: Historical Notes on Some Interactions between Science and Theology* (Vatican City: Vatican Observatory Publications 2007) chap. 3.

[12] *Hexaëmeron*, 2,3 (PL 14,150) in Ref. [11] p. 3:16.

[13] See Ref. [11] chap. 13.

[14] See Ref. [4] pp. 38–39.

[15] *Apologia*, 1,59.

[16] *Adversus Haereses*, 2,1.1; 3,8.3.

[17] *Stromata*.

[18] See Ref. [17] II,1,1.

[19] According to H. Chadwick, the first suggestions concerning the *creatio ex nihilo* appeared in the writings of the Egyptian Gnostic, Basilides; for more see: H. Chadwick, *Freedom and Necessity in Early Christian Thought about God*, Concilium 166, 1983, p. 11–13 (English version).

[20] *De Principiis*, 1,3.3; 2,1.4.

[21] For more about Augustine's seed-principles see: E.McMullin, Introduction "Evolution and Creation", in: *Evolution and Creation*, ed.: E. McMullin (Notre Dame, IN: University of Notre Dame Press 1985) pp. 1–56; especially pp. 4–16.

[22] *De Genesi Contra Manicheos*, 1,22–23.

[23] See Ref. [22] 7, 28.

[24] *De Principiis*, 3,5.3 (SC 268,222).

[25] *Confessiones*, 11,10.

[26] *De Civitate Dei*, XI, 5.

[27] *Confessiones*, 11,14.

[28] *De Principiis*, 3,15.

[29] *City of God*, Book XI, ch. 6, translated by H. Bettenson (London: Penguin Books 1972).

[30] *De Principiis*, 3,3,4–5.

[31] *De Civitate Dei*, XII,13 (PL 41,362).

[32] See Ref. [4] p. 29.

[33] See Ref. [4] p. 30.

[34] See Ref. [4] p. 32.

[35] See Ref. [21] pp. 1–2.

CHAPTER 8

[1] M. Heller, *The New Physics and a New Theology* (Vatican City: Vatican Observatory Publications 1996) chap. 2.

[2] Which can be translated as "Yes and No," or "For and Against," or "This Way and That," see: W.A. Wallace, "The Philosophical Setting of Medieval Science," in: *Science in the Middle Ages*, ed.: D.C. Lindberg (Chicago: University of Chicago Press 1978) p. 94.

[3] P. Kibre and N. Siruisi, "The Institutional Setting: The Universities", in Ref. [2] p. 120.

[4] See Ref. [2] p. 97.

[5] See Ref. [2] p. 104.

[6] E. Grant, "Science and Theology in the Middle Ages", in: *God and Nature*, eds.: D.C. Lindberg and R. L. Numbers (Berkeley – Los Angeles – London: University of California Press 1986) p. 54.

[7] See Ref. [2] pp. 108–109.

[8] See Ref. [7].

[9] We have borrowed the following analysis from: C.S. Lewis, *The Discarded Image* (Cambridge, UK – New York: Cambridge University Press 1988) chap. 1. Although

he explained "the medieval situation" in general, his analysis applies in particular to the origin of the Scholastic method.

10 See Ref. [9] p. 5.

11 See Ref. [9] p. 10.

12 See Ref. [9] p. 11.

13 This is despite the fact that in doing "School philosophy'" medieval masters (following St. Thomas) carefully distinguished what could be called the "God of philosophers" and the "God of theologians". This distinction reflects the difference between philosophy, which is based only on natural reason, and theology, which also takes into account revealed truths.

14 An important study in this field is: A. Funkenstein, *Theology and the Scientific Imagination from the Middle Ages to the Seventeenth Century* (Princeton: Princeton University Press 1986).

15 See Ref. [14] p. 122.

16 See Ref. [14] p. 123.

CHAPTER 9

1 Physics, book VI, 239b; translated by R. P. Hardie and R. K. Gaye, *The Internet Classic Archive*: http://classics.mit.edu

2 There exists a fourth antinomy, called "Stadium". It is more complicated, and has been transmitted to us by Aristotle in an unclear form. We shall not discuss it.

3 C.B. Boyer, *The History of the Calculus and Its Conceptual Development* (New York: Dover Publications 1959) chap. 2.

4 G.J. Whitrow, *The Natural Philosophy of Time*, second enlarged edition (Oxford: Clarendon Press 1980) pp. 190–200.

5 He did that in the works: *Essai sur les données immédiates de la conscience* (Paris, 1889); *L'évolution créatrice* (Paris, 1907).

6 See Ref. [4] chap. 2, pp. 6–7.

7 Some people claim that it is only nonstandard analysis, invented by Abraham Robinson, that solves the Zeno paradoxes. In this analysis the so-called nonstandard numbers exist which can be interpreted as "infinitesimally small." We should remember that the nonstandard analysis, as a purely mathematical theory, is not able to solve anything referring to the real world (for the same reason as the standard analysis is unable to do so). However, it is possible to resolve Zeno's paradoxes by combining nonstandard analysis with the method of mathematical modeling. It happens quite often that the same physical phenomenon has more than one mathematical model.

CHAPTER 10

[1] Physics, Book III, 201a; translated by R.P. Hardie and R.K. Gaye, *Internet Classics Archive*, classics.mit.edu

[2] See Ref. [1].

[3] Aristotle's original text runs as follows: If A "has moved B a distance G in a time D, then in the same time the same force A will move B/2 twice the distance G, and in D/2 it will move B/2 the whole distance for G: thus the rules of proportion will be observed" (Book VII, 250a).

[4] This interpretation was proposed in: S. Toulmin and J. Goodfield, *The Fabric of the Heavens* (New York: Harper and Row 1961) pp. 97–101.

[5] Physics, Book VIII, 215b. This particular formulation appears in a longer passage in which Aristotle analyses the causes of motion.

[6] Physics, Book IV, 215a.

[7] See Ref. [6].

CHAPTER 11

[1] S. Drake, *Galileo Studies* (Ann Arbor: The University of Michigan Press 1970) p. 26.

[2] *Études d'histoire de la pensée scientifique* (Paris: Gallimard 1973) p. 117.

[3] See Ref. [1] p. 26.

[4] Ptolemy, "The Almagest", Book 1, in: *Great Books of the Western World*, vol. 16, ed.: R. M. Hutchins (Chicago – London – Toronto: W. Benton 1978) p. 12.

[5] Nicolaus Copernicus, "On the Revolutions of the Heavenly Spheres", Book 1, in: *Great Books of the Western World* (New York – London: Encyclopedia Britannica, Inc. 2007) p. 518.

[6] Galileo Galilei, *Dialogue Concerning the Two Chief World Systems*, translated by S. Drake (Berkeley – Los Angeles: University of California Press 1953) p. 114.

[7] See Ref. [6] p. 116.

[8] A. Koyré, *Études galiléennes III: Galilée et la loi d'inertie* (Paris: Éd. Hermann 1939) p. 60.

[9] Quoted from P. Tannary, "Galileo and the Principles of Dynamics," in: *Galileo – Man of Science*, ed. by E. McMullin (New York – London: Basic Books 1967) pp. 164–165.

[10] See Ref. [9] p. 172.

[11] He included in the *Discorsi* his earlier work *De motu locali*.

[12] See Ref. [2] pp. 202 and 264.

CHAPTER 12

1. All quotations from Newton in this section are taken from: *Newton's Principles of Natural Philosophy* (London: Dawsons and Mall 1968).

2. The second definition deals with the "quantity of motion," nowadays called momentum.

3. For completeness, let us also quote the Third Law (although it does not take part in the plot of our book): "To every action there is always opposed an equal reaction; or, the mutual actions of two bodies upon each other are always equal, and directed to contrary parts."

4. G.J. Whitrow: *The Natural Philosophy of Time* (London – Edinburgh: Thomas Nelson and Sons 1961) p. 129.

5. Barrow continues: "not any more than it implies rest; whether things move or are still, whether we sleep or wake, time pursues the even tenour of its way" (Lectiones geometricae, London 1735, p. 35). We can identify here the source of Newton's own definition of the "absolute, true, mathematical time" which "from its own nature flows equably without relation to anything external."

6. See Ref. [4] p. 132.

7. M. Jammer, *Concepts of Mass in Classical and Modern Physics* (Cambridge, MA: Harvard University Press 1961).

8. E. McMullin, *Newton on Matter and Activity* (Notre Dame, IN: University of Notre Dame Press 1978).

9. René Thom contrasted this approach of Newton with the method of Descartes who, in his mechanics of "direct contact" (by friction and collisions), explained everything but calculated nothing; see, R. Thom, *Stabilité structurelle et morphogénèse*, deuxième édition (Paris: Interéditions 1987).

10. "A Scheme for Establishing the Royal Society," in: *Newton's Philosophy of Nature – Selections from His Writtings*, ed.: H. S. Thayer (New York: Hafner Press 1974) pp. 1 and 181.

11. See Ref. [10] p. 7.

12. In this short chapter we were not able to render full justice to Newton's achievement. To at least partially compensate for this shortcoming we refer the reader to the ample literature available on this topic; for instance: *The Cambridge Companion to Newton*, ed.: I. B. Cohen, G. E. Smith (Cambridge, UK: Cambridge University Press 2002); especially the papers: G. E. Smith, *The Methodology of Principia*, pp. 138–173; W. Harper, *Newton's Argument for Universal Gravitation*, pp. 174–201.

CHAPTER 13

[1] Of course, this universe is too simple to admit a system as complex as a physicist; however, we assume this for the sake of the argument.

[2] B. Van Fraassen, *The Scientific Image* (Oxford: Clarendon Press 1980).

[3] Van Fraassen believes that physical theories are neither true nor false, but only empirically adequate or empirically inadequate.

[4] Logical deduction leads always to true conclusions provided the premises are true.

[5] This kind of induction is called incomplete. Induction is complete if the number of cases is finite, and all cases have been investigated.

[6] *Principia*, vol. 2, translated by A. Motte, ed. by F. Cajori (Berkeley – London – Los Angeles: University of California Press 1934) p. 400.

[7] *Objective Knowledge* (Oxford: Clarendon Press 1975) p. 4.

[8] Named after Thomas Bayes (c. 1702–1761), British mathematician and Presbyterian minister.

[9] For further reading we recommend, among others: W. Salmon, *The Foundations of Scientific Inference* (Pittsburgh: University of Pittsburgh Press 1966); B. Skyrms, *Choice and Chance*, 4th ed. (New York: Wadsworth 1999).

[10] K. Popper, *The Logic of Scientific Discovery* (London: Routledge 2002).

CHAPTER 14

[1] *Science and the Modern World* (Glasgow: Collins 1975) p. 98.

[2] This function is essentially the difference between the total kinetic energy of the system and its total potential energy.

[3] Strictly speaking, all vectors in a given Hilbert space pointing in the same direction represent the same state of the considered quantum system.

[4] Title of S. Weinberg's book, *Dreams of a Final Theory* (New York: Pantheon Books 1992).

[5] J. Zycinski, "Status of Ideal Objects and Philosophical Implications of Contemporary Physics" in: *W kregu filozofii Romana Ingardena*, ed. W. Strozewski (Warsaw – Crakow: A. Wegrzecki, Wyd. Naukowe PWN 1995) pp. 97–109, in Polish.

AFTERTHOUGHTS

[1] The full title of Resnik's paper is: "Mathematics as a Science of Patterns: Ontology and Reference" *Nous* 15, 1981, 529–550.

[2] Relation is defined to be a subset of a Cartesian product of sets.

[3] One can construct an axiomatic category theory without mentioning objects in the axioms, but it turns out that objects always implicitly exist, i.e. their existence can be deduced from the axioms.

[4] E. Landry, *Category Theory: The Language of Mathematics*, scistud.umkc.edn/psa98/papers/

[5] J. Worrall, "Structural Realism: The Best of Both Worlds?", *Dialectica* 47, 1989, 97–124.

[6] J. Ladyman, "What is Structural Realism?", *Studies in the History and Philosophy of Science* 29, 1998, 409–424.

SUBJECT INDEX

~

Printed by Books on Demand, Germany